Aristotle to Einstein

*A Historical Development
of the Theory of Relativity*

Introduction

Prior to the 20th century all physics theories assumed space and time to be absolutes... With the development of special and later general relativity theory in the early 20th century, the role of space and time in our theories of physics changed dramatically. Instead of being a passive background, space and time came to be viewed as dynamic actors in physics, capable of being changed by the matter within them and in turn changing the way that matter behaves.

Gary Felder –
Bumps and Wiggles:
An Introduction to General Relativity

Twenty years ago, the website Physics World polled some 400 people, asking them the question, "Which scientists made the most important contributions to physics?"[*] Overwhelmingly, one person consistently ranked highest: Sir Isaac Newton! And, frankly, they would not be wrong. For a moment, consider some of what Isaac Newton accomplished during his lifetime. He was the first to offer a more accurate explanation of gravity and laws of motion. And when the above is combined with his discovery of calculus, Newtonian mechanics made the movement of the very heavens comprehensible. Even today, more than 300 years later, Newtonian mechanics remain the cornerstone of university science and engineering programs. As a result, the construction of modern bridges and dams, the guiding of sophisticated satellites and probes, even our electronic age, all these endeavors, and more, are achieved (all or in part) through innovative ideas conceived by Isaac Newton. What's more, no simpler or more functional alternative is likely to be found. For these reasons, Isaac Newton is rightly regarded as the most influential physicist to date.

---

[*] https://physicsworld.com/a/newton-tops-physicsweb-poll/

3

With that being said, in any list of the world's greatest physicists, you will find Albert Einstein. And yet, ironically, although most could recite (and accept) Newton's laws of motion, most do not fully comprehend Einstein's theories (there are two) of relativity. Indeed, while most know that Einstein's theories are meaningful and immediately recognize *the* equation, they often do not grasp the sheer technical beauty of his theories. Nor do most appreciate the elegance with which they answer many fundamental questions of *how* the universe behaves. This book seeks to remedy that, to provide a simple explanation of Einstein's theories of relativity. It will achieve this through a historical perspective; it will follow the theories' development as these were logically derived. Furthermore, contrary to what many have supposed, this explanation can be achieved without a great deal of advanced mathematics. How is that possible?

Many who tend to shy away from science seem to have one thing in common, their mutual dislike of mathematics. They seem to believe that any understanding of science is inseparable from understanding advanced mathematics. Yet this is, frankly, not true. Admittedly, there is an undeniable symbiosis between the sciences and math. So, anyone with a fair to moderate knowledge of both will only increase their appreciation for the universe's rationality. Still, it must also be understood that it is possible to grasp the concepts of Einstein's theories of relativity without fully comprehending their underlying mathematics. Why can this be stated with such certainty? For the simple reason that *the physical universe does not operate based on mathematical principles, mathematical principles are based on the universe.*

Consider the above statement another way. There is no universal computer program (at least we've not discovered one) crunching the numbers somewhere behind the scenes to keep the universe operating. No, the physical universe's operation, from the smallest subatomic particle to the most immense galaxy, is based on fundamental laws, not numbers. For instance, heat always flows from the object that is hotter to the cooler one, which is a fundamental law that the universe obeys. Why does this occur? We don't know. But what we do know is that every experiment we've ever conducted verifies that this law *always* holds true. More notably, our knowledge of this fundamental law explains

4

why the handle of a metal spoon in a boiling pot will always heat up. And thus the reason we think twice before grabbing it. So, without opening a single book on thermodynamics or calculus, we already thoroughly grasp a fundamental law of the universe. Yet, this begs the question: If we can understand scientific principles without knowing their applicable mathematics, then of what use is math when studying science? We use mathematics to do two things in science and engineering: (1) to determine how some physical phenomenon has or will progress and (2) to mimic some physical phenomenon. Consider, for instance, that there are 24 hours in a *day*.

Does the fact that there are 24 hours in a day somehow make the number 24 special? Nope. The assigning of 24 hours to a day is merely the *formula* humans have devised to describe the time it takes for the Earth to rotate on its axis: 24 divisions we call *hours*, made up of 60 divisions called *minutes*, that are themselves divided into 60 divisions we call *seconds*. As a result, suppose we were on the other side of the universe, unable even to see the light being emitted from our home star. The above formula would help us always to be able to calculate the exact time in our home city of, let's say, Jakarta. Or suppose we were traveling in the void between the stars. In that case, we could use the formula to mimic the length of a day on our ship – dividing our time into reoccurring 24-hour periods. Even if the formula that described the length of a day was divided into 13 divisions called *khoops*, it wouldn't affect the Earth's rotation on its axis or the mechanisms that cause it. Without a doubt, it is entirely possible to exhaustively describe how the Earth rotates on its axis every 13 khoops instead of every 24 hours. Yes, it is quite possible to understand the theories of relativity without understanding their mathematics. However, to do this will require that we first answer two essential questions: What exactly are Einstein's theories? And why are they called theories of *relativity*?

Why Relativity?

Before the mid-19th century, humanity accepted a straightforward interpretation of reality – we existed in a universe composed of three dimensions: height, length, and width. And this belief was not merely the conceit of our senses. Thousands of years of recorded experience, experimentation, and mathematics confirmed it. Yet, this belief came with very far-reaching implications. For instance, the calculated volume of one cubic meter – whether measured in London, in New Delhi, or 150 million kilometers away on the Sun's surface – was considered constant. A cubic meter was thought to be identical at all velocities and in every location in the universe. Granted, as Newton described, the gravity at the surface of the Sun would be greater than that on the Earth due to the Sun's greater mass. But the Sun's larger gravitational field was not thought to, in any way, affect the measured three dimensions of space. After all, gravity was assumed to be a force like magnetism. Except where magnetism was the attraction between ferromagnetic metals, gravity was the attraction between objects having mass. Space was believed to be constant and absolute, unaffected by the mass that inhabited it.

In a similar vein, time was also thought to be unchangeable. For instance, imagine it is 1835, and our 19th-century counterpart is reading a book while seconds tick away on a nearby clock. At the same instant, in the night sky above, the brilliant streak of Halley's Comet can be seen. Nothing in human experience before this would have suggested that the *rate* at which time progressed in these two locations, or, more precisely, that the inertial time frame at Earth's surface compared to that atop Halley's Comet could be different. But as we shall explore, Albert Einstein would change this inert view of both space and time. He would show that the *shape* of space (i.e., its height, length, and width) could be physically distorted within a strong gravitational field. And that this distortion affected the rate at which time flowed. In other words, both space and time, just as velocity and mass, could be made to vary. Space and time, according to Einstein, were not static elements. Instead, they were dynamic components in continuous flux, yet always perfectly balanced to keep us almost entirely unaware. Yes, space and time are *relative*.

Consider once more the example of our historical counterpart of 1835. When Halley's Comet passed the Earth in the 19th century, it traveled at more than 241 000 km per hour. Now imagine that our counterpart could have ridden atop Halley's Comet. According to special relativity, time would have been moving a bit more slowly than it would have been back on Earth. And this is how the theory gets its name. It is called special relativity because it deals *only* with the *special* case of an object traveling at high speed, without the influence of gravity. As that object's relative velocity (i.e., its motion in comparison to our own) increases, then the rate at which time flows for that object, as well as the object's length in the direction of travel, decreases. Moreover, as the object's velocity increases, its mass also increases. Therefore, according to special relativity, space, time, mass, and energy are all interconnected. Again, consider our 19th-century counterpart.

In 1835, our counterpart would have been familiar with Newton's Second Law of motion (force = mass x acceleration). He would have also understood that the greater an object's acceleration, the greater its force. Thus, a ball accelerated to 145 km per hour hurts more when it strikes you than that same ball thrown at 8 km per hour. However, our counterpart would not have appreciated the true consequences of Newtonian mechanics placing no limit on the speed at which an object could travel. To him, the more energy you supplied an object (in the form of acceleration), the more its velocity continued to increase – without any upper limit. But Einstein changed this view. He would show that as the velocity of an object approaches the speed of light, it would experience a disproportionate increase in its mass. And it is this increase in mass that serves as a cosmic speed limit. To understand how this happens, let's imagine supplying energy to an object to accelerate it at a specific rate.

Let's say you and some friends are taking a joyride in a spaceship, and you feel the need for speed! You push a button, and for 5 seconds, the rocket accelerates by 2 m/s^2. According to Einstein, two things simultaneously occur as you accelerate. First, as expected, a portion of the rocket's energy goes into increasing the ship's velocity: you speed up. However, the remaining portion of the rocket's energy isn't used to increase your velocity. Instead, this latter portion of energy actually increases the ship's mass; the ship (and everyone and thing on it) becomes slightly heavier. As a result, if you push the button a second time, now that the rocket has more mass (i.e., it's heavier), it would take more energy to accelerate it *at the same rate* of 2 m/s^2. The more you accelerate, the faster the rocket's velocity, thus the greater the percentage of energy that goes into increasing the rocket's mass, not its velocity. The consequence of the above process is that objects possessing mass now had a cosmic speed limit that they *could not* reach: the speed of light.

As an object *approaches* the speed of light, more and more of the energy used to accelerate it goes into increasing its mass. And this exponential increase in mass serves to prevent the object from reaching the speed of light; any object possessing mass can only approach but never achieve light speed. Albert Einstein explained this effect in his theory of special relativity, published in 1905. However, Einstein's description of how the universe behaved didn't end there. In 1915, he published his paper on general relativity. Special relativity explained the behavior of light and objects traveling at a uniform velocity below the speed of light and outside the influence of gravity. General relativity sought to describe gravity and the behavior of objects possessing mass *within* a gravitational field.

According to general relativity, the sensation we perceive as a "pulling" force that keeps us so securely anchored to the ground is, in reality, the distortion of space-time. In effect, like a sponge being squeezed on all sides, Earth's mass squeezes the space-time surrounding it, curving it inward toward the Earth's center. Thus, gravity is not a *force* that attracts objects having mass. Gravity is the increased "squeezing" or distortion of space-time that we sense as an acceleration that continuously "pulls" us toward the Earth's center. Furthermore, because of this distortion's shape, massive objects like the Earth and the Sun are spoken of as residing within a *gravity well*. (Figure 1) And just to be clear, these relativistic effects are not just fascinating footnotes found on the pages of science textbooks. Both special and general relativity have a direct, real-world impact on all of us. Indeed, many modern technologies we rely on would not operate as we anticipate if we did not correctly account for relativistic effects. Consider a communications satellite in orbit about the Earth.

To keep a satellite in orbit, engineers must ensure that it maintains an average velocity of 11 200 km per hour. As a result, according to special relativity, the satellite's clock will <u>slow</u> by 7.2 microseconds per day. However, according to *general* relativity, the rate at which time flows increases as the force of gravity

8

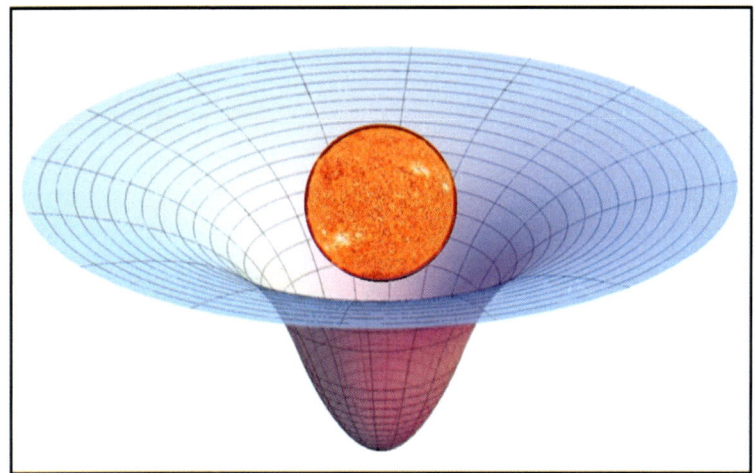

Figure 1. According to general relativity, gravity is not a force like magnetism. It is, instead, the physical distortion of space-time around objects having mass. Thus, the Sun creates a gravity well, the literal inward curving of 3D space directed toward its center.

decreases. So, for a satellite at an altitude of 35 700 km* (i.e., far outside of Earth's gravitational field), the rate of time as measured by its clock <u>increases</u> by 45.9 microseconds per day. The total relativistic effects on the satellite are determined by subtracting the above numbers (45.9 – 7.2). The satellite's clock will run 38.7 microseconds per day faster than a clock on Earth. Yet why would we concern ourselves with 38.7 *micro*seconds? While such a small fraction of time doesn't sound like it would make a big difference, its effects can be catastrophic if not considered. Why?

Well, when a GPS satellite sends its data to receivers in our vehicles, included in that data is the satellite's position. However, for any moving object, its position at any given moment is determined by keeping track of its speed. And herein lies the challenge: speed is a measure of distance divided by *time*. So,

---

* The required altitude of a satellite depends on its type. Some common altitudes for typical satellites are mapping and spy satellites, 480-966 km; scientific satellites, 4800-9600 km; GPS satellites, 9600-19 000 km; communications satellites, 35 700 km (geostationary orbit).

what exactly happens when a satellite's internal clock runs 38.7 microseconds per day faster than clocks operating on the Earth?[*]

The first thing affected is the satellite's speed. Recall that speed is distance divided by *time*. But, if time progresses at a slower rate, the satellite's calculated position will be off by some 152 mm per day. Still, this doesn't sound like much. Yet, if this deviation were not taken into account, in a week, the satellite's position would be off by 1 meter. In a year, 52 meters, and so on for its entire service life. And since there are some 24 different GPS satellites in orbit, all circling in different directions about the Earth, each satellite's location could quickly become lost in the skies above. But this is only half the problem. We use GPS satellites to pinpoint our position on the Earth. However, every 0.1 microseconds of discrepancy will produce a 30-meter error on the Earth's surface. Thus, a 38.7-microsecond disparity equates to more than 11 km at the Earth's surface! The latter time effect is known as time dilation.

As a result of the above, engineers reconcile relativistic effects in two ways. First, weekly updates are performed using the satellites' thruster rockets; this keeps the satellites in their relative positions. Lastly, they slow the onboard clocks of orbiting satellites by approximately 38.7 microseconds per day to keep them relatively synchronized with Earth-based clocks.[†] And yet, this was not the only physical phenomenon that Einstein explained. His general theory of relativity suggested another extraordinary feature of nature.

The theories of relativity also proposed that there could exist stars in the universe that were so massive their gravity could warp the fabric of space and time around them to an infinite degree. As a result, not even light would be able to escape from such stars. And again, according to general relativity, it wouldn't be that the

---

[*] Keep in mind that, although it is easier to speak of the satellite's clock as running faster or slower when compared to clocks on Earth, time itself is what actually has changed. In other words, it is not a mechanical issue, as if the clocks are physically slowing down. Time is moving at a different rate in each location.

[†] An excellent discussion of the GPS system can be found at:
http://gsmserver.com/articles/gps.php

star's gravity was pulling on the light. Instead, the space-time surrounding the star would become so distorted that light would be trapped within the gravity well. The collapsed star would thus emit no light; it would become a *black hole** against dark space.

Another physical effect that relativity explained is that the faster an object travels, especially as it approaches light speed, the shorter it becomes when measured in the direction of travel. This process is known as *relativistic length contraction.* But what's extraordinary is, at these exceptional velocities, you wouldn't even be aware that you, and everything with you, had experienced any length contraction. Indeed, the latter outcome lies at the core of Einstein's theories.

The principle of relativity asserts that the laws of physics are perceived the same for all persons traveling at any *uniform* velocity. Therefore, despite length contraction and time dilation happening around us every day, so long as we are not accelerating, we remain happily unaware that these effects are occurring. Yet, to really appreciate what this implies, consider two persons; the first is inside a large box that is motionless in space. The second person is also inside a box; however, he and his box travel through space at a uniform velocity of more than 362 000 000 km per hour. Relativity tells us that these individuals could conduct no experiment that could conclusively prove which one was at rest and which was in motion. Thus, all the laws of physics would behave the same for both persons.

So time and space are relative – dependent on each person's perspective within his inertial frame of reference (i.e., in his *own* box). Indeed, you can prove this yourself any time you're traveling in your car on the highway. As long as you maintain a uniform velocity, you can pour a hot cup of coffee from one cup to another without it spilling. The coffee will behave as it usually does when you conduct the same experiment while standing at your kitchen counter. You would get the same results even when traveling at more than 800 km per hour in an airplane. The laws of physics will *always* behave the same so long as uniform velocity

---

* Physicist John A. Wheeler (1911-2008) coined the term *black hole* in 1967.

is maintained.* However, time dilation, the warping of space, and black holes were still not the most significant aspects of Einstein's theories of relativity. The most remarkable feature arose in special relativity when Einstein sought to compare energy with mass.

In 1905, as part of the special theory of relativity, Einstein proposed an equation that, even in its mathematical form, was exceptionally profound in meaning. Yet, the equation was stunningly elegant and simple, so simple that few had any difficulty understanding what it implied. The equation showed what is known today as *mass-energy equivalence*, a discovery that instantly fired the imaginations of engineers and scientists alike. They imagined the production of unlimited amounts of energy from small amounts of mass, potentially solving all the world's energy needs. Alternatively, others conceived of producing mass from energy: building machines capable of creating full-course meals by merely being plugged into an electrical outlet. Or perhaps matter-energy transportation and matter-antimatter space-craft. Einstein showed us that these two facets of reality (mass and energy), which we had always treated as separate prior to 1905, were not mutually exclusive. Accordingly, now have started to grasp some of the *intricacies* of Einstein's theories, we're now ready to have them explicitly stated.

Special relativity describes the structure of both space and time in the absence of gravity and is based on two fundamental premises. (1) The speed of light is *always* constant. (2) The *principle of relativity*: that the laws of physics behaves the same for all persons moving at a uniform velocity. From these postulates (and the consequences they imply), we can ascertain the structure and behavior of the universe. For example, there is *time dilation*, meaning that the rate at which time flows is <u>not</u> absolute. There is *relativity of simultaneity*, that the exact timing of two spatially separated events <u>cannot</u> be stated with absolute certainty. There is *length contraction*, that the physical size of an object can

---

* This is not the case when the frame of reference you're in is subject to a positive or negative <u>acceleration</u>. You can prove this by conducting the same coffee experiment, except try to pour that cup of coffee while the driver slams on the brakes. The coffee will not pour straight down into the cup; the stream of hot coffee will pour, if you are fortunate, onto the floor in front of you.

be made to vary based solely on its velocity. And finally, mass-energy equivalence, or $E = mc^2$. As we shall see, all these phenomena follow from the two stated premises above.

General relativity explains the behavior and nature of gravity. It is based on a single notion: the *equivalence principle*, which affirms that it is impossible to distinguish between the acceleration caused by gravity and an equivalent amount of acceleration caused by something else. For example, consider the acceleration of a jet traveling down a runway. The inertia that presses one against the back of their chair is not caused by the acceleration of gravity, which keeps one secured to the ground. However, because the *effects* are similar, it must be accepted that all types of acceleration, whether caused by a gravitational force or by an airplane, must also be equivalent. It is from this postulate that the revolutionary model of general relativity arises. Furthermore, general relativity states that instead of gravity existing as a *force*, as the magnetic or an electric field, gravity is the warping of the fabric of space-time itself, into which objects that have mass *fall*. These are Einstein's theories of relativity.

And so, with no further delay, we return to the purpose of this book: understanding *how* scientists could ever come to believe in a world that behaved so peculiarly. What scientific discoveries made before Einstein would ever lead him to conclude that the universe obeyed these laws of relativity? The question above is what this book will answer. It will do so by stepping back in history and following the theory's development over 2,100 years. It is as much an exploration of human curiosity and ingenuity as a desire to understand the universe. We shall see how each person who contributed to this theory, in their own way, reached new, more accurate conclusions. And in the end, we will better appreciate that it was just as much a philosophical journey as it was scientific.[*]

---

[*] Most of the examples throughout the book are greatly simplified, especially when dealing with the speed of light in the final chapters. This was done to keep the explanations understandable. However, the principles still apply.

Chapter 1

Where to begin? Contrary to popular belief, the history of science is not merely composed of several crucial discoveries arranged like steppingstones on a path. Instead, history is like a tapestry that consists of many threads, with all its strands intricately intertwined. The arts, the rise and fall of civilizations, and the mass migration of ancient peoples (or single person) have all played a role. And to pull on any of these threads would alter the entire fabric. So, where do we begin? As the title suggests, the most direct link toward relativity begins with Aristotle.

Many scientific discoveries were made before the birth of our first contributor to the theory of relativity. Early thinkers whose particular breakthroughs helped lay the foundation for our modern civilization. However, since our goal is to chart the development of Einstein's theory of relativity, one person stands out as deserving to begin with: the Greek philosopher Aristotle. And this can be stated with such certainty because he founded the study of *logic*.

Aristotle (384-322 BCE)

It was Aristotle's uncanny ability to reason, which, upon refinement, became known as the scientific method, that proved to be one of the most powerful tools of science. In fact, as we'll see, Aristotle's early form of logic, although crude, made his scientific theories almost irrefutable; it took 2,000 years to disprove the last of his theories. So, while not belittling the ancient Egyptians' discovery of geometry or the Chinese in their conceiving of the decimal point, our discussion will begin with Aristotle. He represents the most direct line in the development of the theory of relativity. However, before examining Aristotle's methods and ideas, let's first learn a little more about him.

Like many of those we will consider, Aristotle came from a very well-to-do family. He was born in the Greek coastal village of Stagira, on the Thracian peninsula. His mother, born of an aristocratic family from Chalcis of Euboea, was named Phaestis; her family may have even been the village founders. And although it can't be confirmed, it's also possible they were affiliated with the Asclepiadae – a prestigious guild of priests who practiced medicine. However, in Greek culture, the father had the greatest influence on the direction of his son's future.

While history is uncertain of Aristotle's maternal affiliation with the Asclepiadeans, this is not the case with his paternal family line. Aristotle's father, Nicomachus, was the personal physician to King Amyntas III, the father of Macedonian King Phillip, the grandfather to Alexander the Great. Therefore, from birth, Aristotle's social status was all but guaranteed. Indeed, it was precisely for those of high social status that the ancient Greek educational system was established.*

After Aristotle's birth, Nicomachus moved his family to Pella to serve King Amyntas III. So, it was in the Macedonian capital where Aristotle spent his formative years.† In accord with ancient Greek tradition, Aristotle's general education would have been cared for by his mother, at least until he reached the age of seven. At that point, he would have joined one of the local gymnasiums where physical training, along with education in music and poetry, was emphasized. At the same time, Aristotle's primary education would have been cared for by his father. From infancy, Nicomachus would have taught his son biology, preparing him to become an Asclepiadean priest and physician, just as he was. However, even with all these social advantages, an unexpected tragedy visited Aristotle when his mother and father died. It was this misfortune, at such a seminal age, that immediately altered the direction of Aristotle's life.

It's unknown how old Aristotle was when his parents died, though most historians believe he was around 10. What is known

---

* Education in ancient Greece was tailored specifically toward boys from wealthy and influential families. Slaves were never formally educated, and neither were women. The exception was the women of Sparta, who were expected to run the city while the men were away fighting.

† Aristotle is believed to have had two siblings: an older sister, Arimneste, and a younger brother, Arimnestus.

is that, after their passing, Aristotle came under the guardianship of a man named Proxenus, who some suggest was his paternal uncle. In any case, Proxenus was married to Aristotle's older sister, Arimneste, and they were the ones who finished rearing the young Aristotle. Now, it appears that Proxenus was not associated with the Asclepiadeans since, had he been a physician, Aristotle's training as one would likely have continued. Instead, when Aristotle went to Athens in 367 BCE to further his education at one of the academies, his area of study was philosophy, not medicine. But it was in this field where Aristotle's brilliance blossomed, particularly since he studied under the renowned philosopher Plato.

Aristotle remained at the academy for 20 years, most of that time as a teacher. Yet, unfortunately, despite their shared love of knowledge, history tells us that Plato and Aristotle never grew to like one another beyond their being kindred intellectuals, at best. So, it's not surprising that in 348 BCE, after Plato's death, Aristotle was unceremoniously bypassed as the academy's next headmaster. It was instead given to Plato's nephew, Speusippus. Feeling slighted, Aristotle left Athens a year later with a few of his students and friends. He spent the next four years traveling throughout the eastern Mediterranean and Asia Minor.* His sabbatical only ended when King Philip, with whom he grew up, asked him to return to Pella to tutor his 13-year-old son Alexander. An assignment Aristotle eagerly accepted.

Aristotle tutored Alexander for six years until, in 336 BCE, King Phillip was assassinated, and the 20-year-old Alexander became king of Macedonia. And this presented Aristotle with a long-awaited opportunity. Now, with a powerful ally on the throne and ample funds at his disposal, he chose to return to Athens. However, he would not do so as merely a teacher. In 335 BCE, at the age of 50, Aristotle founded a rival academy in Athens called the Lyceum. It would be here that his name would become world-

---

* There is an alternate reason for Aristotle's departure from Athens. Some say that he was compelled to depart after Olynthus, a Greek city near Aristotle's hometown, rebelled against King Philip. The resulting conflict resulted in increased anti-Macedonian sentiment in Athens. King Philip razed Olynthus in 348 BCE.

renowned. At his academy, he pioneered a unique teaching method akin to what we would find today in a university graduate program. This teaching methodology focused on cooperative research and included comprehensive student input and historical and scientific research projects. Over the next 12 years, Aristotle would advance most of the philosophical ideas that he is today celebrated, the same ideas that started humanity on the road to relativity. And it all began with logic!

Aristotelian Syllogism

As already stated, the essential contribution to science that Aristotle made was his development of the first known scientific method. These are the rules that one would follow when analyzing a phenomenon. To appreciate why these rules were so revolutionary, let's take a moment to examine the modern-day scientific method, which consists of 4 primary steps. For example, it begins by asking a specific question:

1. "Why is the sky blue during the day but red at sunset?"

Step 2 is to hypothesize some reasonable explanation that you believe answers the question; in essence, you make an educated guess. So, after doing some research on diffusion, you put forward the following hypothesis:

2. "We know that white light is composed of all the colors of the rainbow. And we know that colors differ due to their wavelength. Therefore, perhaps the sky is blue during the day, but red at sunset, because of their different wavelengths. And, depending on how much atmosphere the sunlight is passing through, the two wavelengths will scatter differently at different times of the day."

Step 3 is to devise an experiment to prove whether your hypothesis is correct.

3. Add a few tablespoons of milk to a large glass of water until the water is a bit murky. Next, shine white light through the liquid.

The final step involves analyzing the data to confirm your hypothesis. If it does not confirm your hypothesis, you use the experiment's data to form a new one.

4. When the above experiment is performed, the following results will occur: Although the light entering the glass is white, the light that radiates from the container's center will be blue. The shorter-wavelength blue light has traveled as far as possible through the liquid and is now scattered. Whereas the light exiting the glass at its opposite end is red, the longer-wavelength red light can travel farther through the medium before being scattered. (Figure 1-1a) Similarly, when sunlight enters the Earth's atmosphere during the day, the first wavelength of light to scatter is blue. However, at sunset, the light must travel farther through the atmosphere to reach the Earth's surface. With the blue light already scattered, all that remains is the longer-wavelength red light. (Figure 1-1b)

The critical point to take away from this experiment is its methodology; it is a logical, repeatable sequence of steps. Such a method is how scientists today go about examining physical phenomena. And although these steps seem reasonable in our modern age, the scientific method has not always existed. As a matter of fact, before the 17th century and Sir Francis Bacon, although scientists made many hypotheses, they conducted few (if any) experiments to verify their conclusions. Even Aristotle's rudimentary scientific method did not include experimentation. Still, despite this shortcoming, Aristotle's method was so groundbreaking that it might have been the underlying cause of the tension between him and Plato. To understand why, let's compare their different approaches to science.

Plato believed that everything on the Earth, including humans, were only imperfect imitations of what existed in the heavens. Furthermore, since humans were bound to the Earth, all they could ever examine was the equally imperfect world around them. Therefore, Plato thought it impossible for imperfect humans, studying an imperfect world, to ever understand the true nature of reality. However, this was not what Aristotle thought.

19

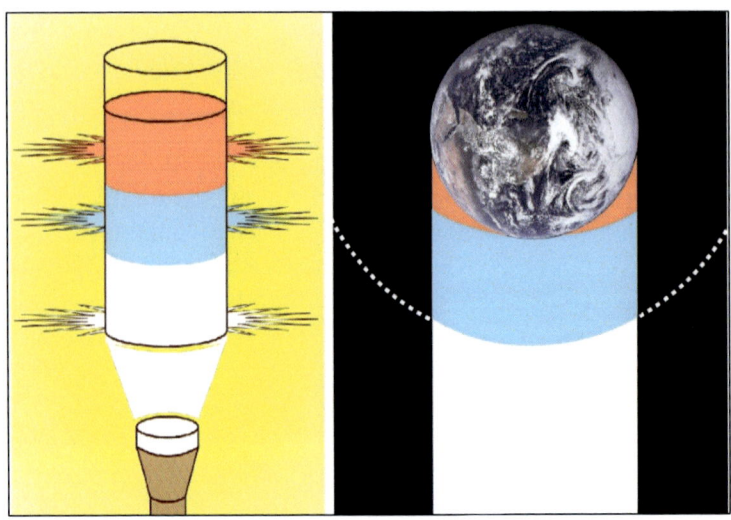

In our experiment, **Figure 1-1a (left)**, white light enters the bottom of the container. As it travels through the water, the color changes – from white to blue, and finally, red. This change occurs because the shorter-wavelength blue light is the first color to be scattered. The longer-wavelength red light can travel farther through the medium before being scattered. This is the same reason that the sky is blue during the day, **Figure 1-1b (right)**. Blue light is the first color to scatter in the Earth's atmosphere. As a result, as the Sun begins to set – and the shorter-wavelength blue light has been removed – only the longer-wavelength red light remains.

Aristotle readily conceded that the physical world was imperfect and impure. He also agreed that the realm of the heavens was, contrastingly, perfect and pure. But where the two men differed was on the degree to which humans could comprehend reality. Could imperfect humans, using nothing but their physical senses, comprehend not only the Earth but even the heavens above? Aristotle answered, "yes!" And without a doubt, that answer would have shocked any traditional-minded, deity-fearing Greek of the era, including Plato. But to fully grasp the core of their dispute, consider a scenario.

A spaceship from another planet crashes in the cornfield of a modest, modern-day farmer. The craft is extraordinarily advanced and irreparably damaged, so much so that the farmer can barely begin to identify the wreckage. And given that the farmer is having a hard time understanding the spacecraft, how could he ever hope to understand the advanced civilization that built it? According to Plato, the spacecraft's technology is too far advanced, the damage is too great, and the farmer is too *simple*. However, despite all these reasons, Aristotle viewed the situation differently.

Aristotle felt that, by virtue of the fact that the aliens existed, there would have to be commonalities between them and the farmer. Thus, as the modest farmer studied the debris, he would inevitably gain a greater knowledge of it and the persons who had constructed it. For example, while carefully examining the debris, the farmer identified a badly mangled chair. In doing so, he would know something more about the aliens, such as their approximate height. He could also deduce that, just as his tractor – a vehicle – has a chair, the wreckage was also, in fact, some type of vehicle. In this way, given time and effort, the more the farmer studied the wreckage, the better his understanding of it and the aliens who constructed it. In essence, Aristotle was an advocate of deductive reasoning: the ability to infer the *specific* by examining the *general*.

By studying the objects that inhabited the imperfect world, Aristotle believed humans could ultimately understand even the *perfect* that existed in the heavens. And this method of drawing conclusions was just the beginning. Because the most significant aspect of Aristotle's approach arose when he combined deductive reasoning with rules that he felt should *always* be followed when examining physical phenomena. It was Aristotle's rules for deductive reasoning that, in the end, set his theories apart from all those that came before him.

A syllogism is discourse in which, certain things being stated, something other than what is stated follows of necessity from their being so. I mean by the last phrase that they produce the consequence, and by this, that no further term is required from without in order to make the consequence necessary.

Aristotle –
Prior Analytics: Part I

To simplify the above phraseology: syllogism (deductive reasoning), as defined by Aristotle, is a way of reasoning on a subject. It allows one to reach conclusions by stating a known (major) premise, then a second (minor) premise. Finally, one draws a conclusion based on how these separate premises relate to one another. For instance:

➢ **Major Premise** – All men are mortal.

➢ **Minor Premise** – Socrates and Plato are men.

➢ **Conclusion** – Therefore, Socrates and Plato are mortal.

From the above examples, it becomes easy to ascertain why Aristotle's theories were influential. A natural consequence of his deductive reasoning was that it was linear, not circular; the premises didn't depend on the conclusion and vice versa. To see why linear logic is always preferred, let's consider an example of one inherent flaw found in a belief of the ancient Greeks.

The pre-Socratic philosopher Empedocles (495–435 BCE) was the first to propose that the universe was made up of five fundamental elements: earth, water, fire, air, and the aether.[*] Plato next theorized that each of the first four elements had associated with them a geometric shape. The element earth was the cube; water, the icosahedron; air, the octahedron; and fire, the tetrahedron.[†] And once geometric shapes had been assigned to the elements, it became possible to explain how more complex elements formed. For instance, water was thought to be composed of two parts air and one part fire. In other words, water, an icosahedron (20 sides), was made up of two octahedrons ($8 + 8 = 16$) and one tetrahedron (four sides). Plato was applying geometry to his theory of matter, which was forward-thinking. According to modern chemistry, the molecular structure of a compound does play a role in defining its chemical makeup. However, there was a

---

[*] According to the Greeks, the aether was the material that filled the region of the universe above the Earth.

[†] An icosahedron (*icos-* is derived from the Greek word for "twenty") is a polyhedron with 20 identical, equilateral triangular faces. An octahedral is a polyhedron with 8 identical, equilateral triangular faces. And a tetrahedron is a polyhedron with 4 identical, equilateral triangular faces.

severe flaw in how Plato arrived at his conclusion, one we can better grasp by examining Plato's primitive molecular theory using Aristotle's syllogism.

Plato's major premise was Empedocles' proposition that the physical realm was composed of the first four fundamental elements. He then associated each element with a specific geometric shape. Finally, the geometric shape determined how more complex elements formed. And here's the problem: Not only was Plato's logic circular, but he also offered no minor premise to reach his conclusions. So the four fundamental elements behaved as they did because of their relationship to geometric shapes (major premise). But with no reason given, he concluded that those geometric shapes were properties of the four elements. So, at best, in scientific terms, Plato's theory was just an unfounded statement, no more valid than saying that each element is linked to a specific letter of the alphabet and, therefore, you could assemble elements like words in a sentence. And why can you assemble elements like words? Because each element is linked to a letter in the alphabet. The logic is circular.

In the final analysis, Plato's theories could never be debated or tested because they were just a compilation of various opinions. Given what history tells us about Aristotle, this form of reasoning would have certainly caused contention between him and Plato. It may have even been the primary cause behind Aristotle developing a scientific method in the first place. Methods that, when applied, made his arguments seemingly irrefutable for centuries. So, let's examine some of these theories of Aristotle.

The word "physics" comes to us through Latin (*physica*), where it means "natural science." Latin got the word from the Greek (φυσική), where it meant "knowledge of nature." *Physics* was the title of Aristotle's 6th treatise, wherein he discussed nature and natural phenomena. And the two aspects of physics that Aristotle discussed that are of immediate interest to us are his theory of matter and his primitive laws of motion.

Aristotelian Physics

Like Plato, Aristotle also began with Empedocles' theory of matter, that the physical world was composed of four fundamental

23

elements. However, Aristotle was careful not to define these elements by linking them to <u>randomly</u> selected geometric shapes. Instead, using his deductive reasoning, he linked them to known sensations. For example, he linked earth with dry and cold; water to cold and wet; air to wet and hot; and, finally, fire to hot and dry. (Figure 1-2) Yes, Aristotle honestly believed that the physical world could be wholly understood through the physical senses. What's more, the selected sensations were not chosen at random, as Plato's geometric shapes were. Fire is hot and dry, and water is normally cold and wet.

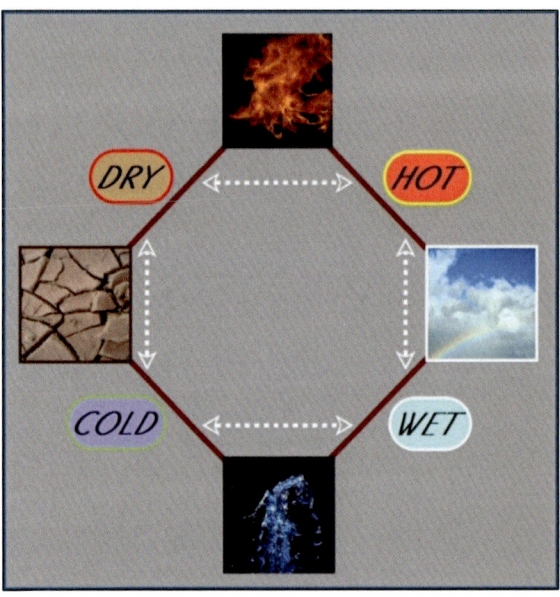

Figure 1-2. Aristotle believed that all matter on the Earth was composed of differing proportions of one or more of the four fundamental elements: earth, water, fire, and air. In turn, the four fundamental elements were the combination of at least two of four sensations: dry, hot, wet, and cold.

Aristotle's major premise was that complex objects were composed of some combination of the four fundamental elements. His minor premise was that each fundamental element was associated with unique physical sensations. Conclusion: the complex objects must share the unique sensations found in their constituent elements. And to demonstrate the validity of his argument, Aristotle offered an example.

According to Aristotle's theory, trees were composed of the following elements: fire, air, and earth. Thus, if his deductive logic was correct, every tree contained these elements, including their associated sensations. And how could he prove his theory right? He gave the example of a burning log, a process that every person in the ancient world would have been all too familiar with.

Imagine an ancient Greek mother cooking a meal for her family over burning logs. The fact that the logs burned would be proof that it *contained* fire, and the fire was dry and hot. The smoke that ascended was the "air" the log contained. And finally, the ash that remained was the "earth," which, after cooling, was dry and cold. Therefore, Aristotle's conclusion seemed proven; trees were composed of three of the four fundamental elements: fire, air, and earth. And with just this single example, it's easily seen how far superior his rules of deduction were compared to any that came before, including Plato's. But Aristotle had only just begun. The power of his deductive reasoning would next tackle a question that no one before him had ever successfully answered; Aristotle sought to explain why objects moved.

The philosopher Zeno of Elea lived around 100 years before Aristotle. This Greek thinker is best known for proposing what today is called "Zeno's Paradox." Stated simply, it goes something like this: A turtle challenges a man to a foot race. Confident that he can outrun a turtle, the man not only accepts the challenge but, when the race begins, he gives the turtle a 10-meter head start. The man then starts to run and, after a few seconds, he's crossed 5 meters, but he immediately encounters a problem. Although he's covered half the distance, he still has five more meters between him and the turtle. A second later, he's covered 2.5 meters, yet he still has 2.5 meters ahead of him. In fact, there is an infinite number of 'half distances' between him and the turtle. Indeed, no matter how fast the man runs, he'll always have half the distance to cover; thus, he will never be able to catch the turtle! This argument is Zeno's Paradox. And although intuitively we know that the man will pass the turtle, no one could explain why. So Aristotle decided to apply his deductive logic to the problem. And in trying to explain Zeno's Paradox, Aristotle devised the first universally accepted *Laws of Motion*.

In devising laws of motion, the immediate challenge was explaining why objects moved at all; more specifically, why objects fell. And what made this question even more difficult for persons of the ancient world was that they didn't fully understand gravity. But Aristotle met the challenge. Building on the idea that objects were composed of the four fundamental elements, to explain gravity, he suggested that all objects were attracted to the *things* that they were composed. This solution was clever, and we can see why by looking again at the example of the burning log.

When a log is burned, the smoke rises into the sky. Why? Because it (the smoke) is composed of air. The ash that remains on the ground does so because it's composed of earth. The same logic can be used when describing the motion of rain. Rain is the element "water." And after falling from the sky, it always pools into rivers, and rivers flow to the ocean. However, a stone that's thrown into the ocean will sink; it moves toward the ocean floor because it's composed of earth. Aristotle named this type of motion *natural* motion.

In contrast, any motion that went contrary to natural motion, Aristotle called *forced* motion. For instance, consider the act of lifting a stone off the ground. Since a stone naturally tends toward the ground, moving it away from the ground was deemed unnatural, thus forced.* Again, his logic was powerfully linear. And as a result, although having no concept of gravity's nature, Aristotle offered a simple explanation of why objects fell; this was his third law of motion. We summarize Aristotle's three laws of motion below.

---

* An important point to keep in mind regarding any new theory is that necessary consequences logically follow. The consequences are what reveal if the theory is correct or not. For example, to explain lightning, it could be theorized that air is composed of metal. One of the consequences of this being that metallic air could conduct electricity. However, other consequences of this theory would be that light from the Sun would not reach the Earth's surface, nor would we be able to breathe. And since both of the latter phenomena do occur, we would have to reject the theory that air is composed of metal. We would then have to formulate a new theory, one that resolved all the above-stated consequences. Therefore, for a theory to be considered valid, its consequences must agree with what we see in nature.

Aristotle's Laws of Motion

1. Nothing moves unless it is pushed.

2. Speed is proportional to the force applied to the object being pushed and inversely proportional to the medium through which it moves.

3. There are three types of motion in the universe. On the Earth, there is *natural* and *forced* motion. The final type of motion is the "perfect" (*circular*) motion represented by the apparent movement of celestial objects.

Another idea that most ancient Greeks believed in was the geocentric model of the universe. (Figure 1-3) This model suggested that the heavens were composed of 56 crystal spheres. Embedded into each sphere were different celestial objects (i.e., various stars and planets). And the orbital period of the particular celestial object determined which sphere it was embedded. Finally, the Earth was at the center of this model, and it sat motionless as the spheres containing the stars and planets rotated around it. Yet, there arose more than a few serious consequences from this model of the universe.

For example, placing the Earth at the center of the universe implied that all motion was absolute (i.e., every object's trajectory could be *absolutely* determined in relation to the Earth). Furthermore, since motion is defined as distance divided by *time*, advocating the geocentric model also inferred that time was absolute. But the consequence of absolute space and time went far beyond the ancient Greeks' understanding, something we will revisit later. Still, for Aristotle, the geocentric model created another, more pressing issue for his newly stated laws of motion.

The idea that the Earth sat at the center of the universe also implied that the Earth was motionless (i.e., it did not rotate). The logic used to reach this conclusion was as follows. If the Earth were rotating, any object dropped (once it left a person's hand) would *not* fall directly to the ground. Instead, while the object was in freefall (thus not in contact with the rotating Earth), it would fall *behind* the person who dropped it. It would be like dropping a rock from the window of a moving car.

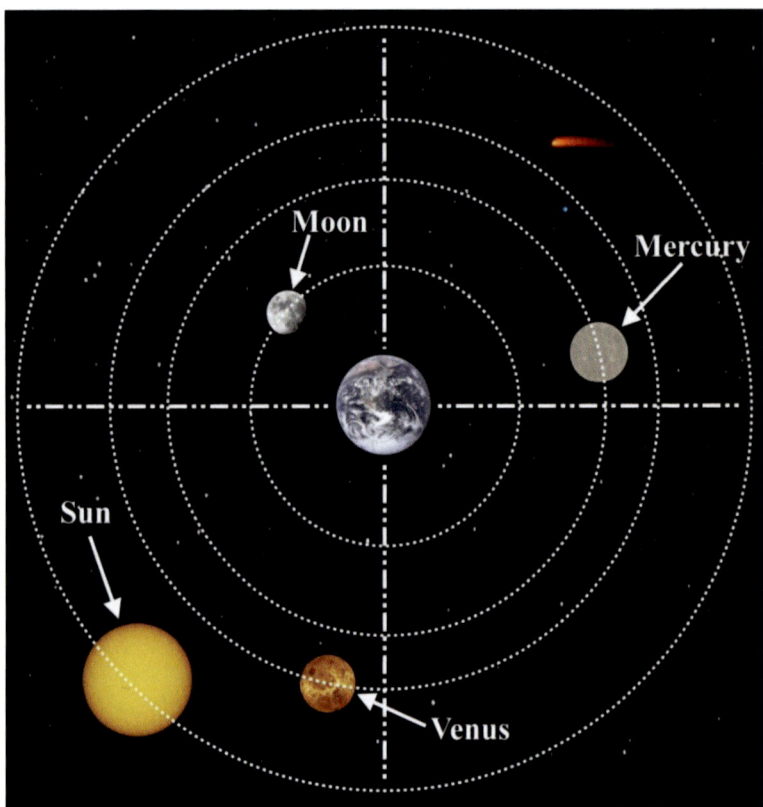

Figure 1-3. According to the geocentric model, the Earth sat at the center of the universe, and everything revolved around it. A natural consequence of this model was that everything in the universe had an absolute position and velocity (as measured from Earth). And, since the Earth sat motionless at the center of the universe, Aristotle also assumed that the natural state of all objects was one of rest.

Imagine holding a rock out of the window of an *accelerating* car. Why does the rock continue at the same rate of acceleration, though not inside the car? Well, so long as the car pushes you forward and you have a firm grip on the rock, the rock will maintain the same acceleration as the car. But what happens when you release the rock? Gravity immediately pulls it toward the ground. And since the car is still accelerating, but the rock is no longer in your hand, it will fall to the ground <u>behind the car</u>. Aristotle used this same logic to prove that the Earth was not in motion. And it agreed with his first stated law: An object that is no longer being pushed (because it is in freefall) will fall to the ground, returning to a state of rest. Finally, Aristotle believed that heavy objects fell faster than lighter ones, such as a brick falling faster than a feather. Aristotle believed this occurred because the medium (i.e., the air) affected the feather more than the brick as they fell. The latter was his second law of motion.

The last concept that Aristotle had to address had to do with *what* "space" (what he called "place") was, which turned out to be problematic.

> *The question, 'what is place?' presents many difficulties. An examination of all the relevant facts seems to lead to divergent conclusions. Moreover, we have inherited nothing from previous thinkers, whether in the way of a statement of difficulties or of a solution... These considerations then would lead us to suppose that place is something distinct from bodies, and that every sensible body is in place.*

Aristotle –

Physics: Part IV

After considering the matter at great length, Aristotle concluded that:

> *By asking these questions, then, we must raise the whole problem about place – not only as to what it is, but even whether there is such a thing.*

Aristotle –

Physics: Part IV

Simply put, Aristotle had no idea of how to define space. Nevertheless, there's one thing of interest to note. Except for energy, each of the areas that Aristotle discussed or implied (i.e., matter, motion, space, and time) were the same things Einstein discussed in his theories of relativity. Yes, Aristotle appears to have been the first to realize that these separate physical

29

phenomena were somehow interconnected. Furthermore, even though his theories were crude compared to what we know today, they still possess an intricate subtlety. But, as we might expect, there were also inherent problems.

Aristotle's first law of motion (nothing moves unless pushed) seems to be a self-evident truth, something that even someone today might believe. In reality, though, this law implied that the natural state of all objects was one of rest. Therefore, any object put in motion because it was pushed will, eventually, stop. And once at rest, it will remain at rest until it is pushed again. This first law appears to agree with what takes place around. Someone rolls a tire, but in due course, the rolling tire will stop; a thrown stone will eventually come to rest on the ground; wind-swept leaves will, sooner or later, settle. And unless someone or something causes these objects to move again, they will remain motionless indefinitely. Hence, Aristotle's first law appeared plausible. Yet, beneath this veneer, there was a crucial flaw.

Imagine that an archer is aiming his arrow at a distant target. Before he releases the bowstring, the arrow is stationary. Why? Because, according to Aristotle, the bowstring has yet to *push* the arrow forward. And, sure enough, the moment the archer releases the bowstring, it begins to push the arrow. As a result, the arrow flies off, but this is where the problem arises. Remember, the ancient Greeks had no concept of inertia (i.e., that an object in motion *wants to stay in motion*). So, the question becomes: if an object *only* moves when pushed, why do arrows continue to fly toward their target? This question was the first that Aristotle had to answer.

Let's consider another problem Aristotle faced. His second law of motion was that speed was proportional to the force applied to the object being moved and inversely proportional to the medium through which it moved. Interestingly, this law applied an algebraic formula to an object's motion, perhaps the first in history to do so. If we were to write his law out in modern mathematical notation, it might appear as follows:

$$v = f/R$$

This formula states that the velocity (v) of a pushed object was directly related to the applied force (f), divided by the resistance (R) of the medium through which the object moves. The significance of what Aristotle was here suggesting lies with the variable R.

Aristotle wanted to explain, for example, why a rock falling in water descended more slowly than a rock falling through the air. His solution was that the rock fell more slowly through water because it (the water) is denser than air and, thus, offers more resistance (R), resulting in the rock falling more slowly. This explanation was reasonable and straightforward and seemed to explain why heavier objects fell faster than lighter ones – whether in air or water. Again, his conclusions had the appearance of agreeing with everyday experience. But there remained a problem, consider once more the falling rock.

Instead of water, imagine dropping a rock from the peak of Mount Everest. According to Aristotle's theories, since the air at that altitude is much thinner, the rock would fall *faster* than the same rock at sea level, where the air is denser. But this creates a paradox. If you follow this mental exercise to its conclusion, it would mean that this same rock, if dropped in a vacuum (where there is *no* air resistance), would fall at an infinite speed. Even for someone not living in our scientific era, it was impossible to imagine an object falling at an *infinite* velocity. So, how could Aristotle solve this falling rock problem and that of the flying arrow? The answer he suggested, though unequivocally wrong, was imaginative. It also demonstrated just how deeply he thought about the issue.

So, Aristotle refused to believe that an object in a vacuum could fall at an infinite velocity. Thus, the only conclusion he thought rational was that it must be impossible for a vacuum to exist. Therefore, if a vacuum couldn't exist, then a falling object will always experience some sort of resistance to prevent it from falling at an infinite speed. But, in a clever twist, Aristotle also used the impossibility of a vacuum's existence to solve the first problem, that of an arrow continuing to fly when it was no longer being pushed.

Aristotle explained that an arrow, as it traveled toward its target, was displacing air. It logically followed that the displaced air created a partial vacuum in the arrow's wake. Thus, as the air rushed in to fill that vacuum, it (the air) was imparting the *pushing* force that kept the arrow flying. And this was Aristotle's explanation of how an arrow continued to travel toward its target when it was no longer being pushed. It was an elegant solution to a complicated problem and is one reason why, even after 23 centuries, he is still reckoned among history's greatest thinkers.

So, by 322 BCE, it appeared that Aristotle's laws of motion answered all questions in regard to why and how objects on Earth moved. Indeed, even as we read his explanations more than 2,000 years later, they still have a certain sensibleness that seems to make them plausible. However, while Aristotle's deductive logic neatly explained how things worked on Earth, he ran into a more serious problem when he attempted to describe the motion of celestial objects. Aristotle's laws of motion did not explain the movement of the stars and planets. And it was this failure of his theories that indicated that new laws of motion were needed. Let's consider why the movement of celestial objects was so confusing.

Like many in his day, Aristotle thought that the motion of celestial objects, such as the stars, was perfect. They reached this conclusion based on their idea that the circle was a "perfect" geometric figure. And, as viewed from the Earth, the stars do track circular paths in the night sky. But the same cannot be said of the planets.

The origin of the English word *planet* is from ancient Greek, in which it is directly translated as *wanderer*. For instance, if you were to go outside every evening for two years and track the planet Mars, you would notice something peculiar. Instead of Mars consistently traveling in an easterly direction, it at times appears to travel backward. Then, without explanation, this regressive motion halts, and the planet will once again begin to travel from west to east. So, while Aristotle's laws of motion for objects on Earth remained unchallenged (initially), their inability to explain the movement of celestial objects could not be ignored. But the world would have to wait more than 400 years before the next contributor toward the theories of relativity could offer any plausible-sounding solution.

Claudius Ptolemy (c. 90-170 CE)

The peak of the Roman Empire occurred during the 84-year contiguous reigns of emperors Nerva (96–98 CE), Trajan (98–117), Hadrian (117–138), Antoninus Pius (138–161), and Marcus Aurelius Antoninus (161–180). This period in Roman history was characterized by rather good governance and a bolstering of the empire's borders. At the same time, the Roman Empire had expanded to its largest territorial extent. It covered around nine million square kilometers – controlling north-western Europe, North Africa, and the Middle East. Accordingly, this period of prosperity in Roman history is known as "the Five Good Emperors." And yet, other challenges remained.

From 165 CE to 180 CE, the Roman Empire suffered one of its worse outbreaks of pestilence. From the descriptions given, it's believed the disease they endured was either smallpox or measles, the former being most likely. It's estimated that almost 1,000 people *per day* died for the plague's duration. It even took the life of the last of the five good emperors; thus, the reason it's alternately called the *Antonine Plague*. However, in terms of scientific advancements, this was also the period during which two of the most notable 2nd-century CE western scientists lived, the first being the physician Galen of Pergamum.

Galen's medical contributions include his detailed descriptions of human anatomy and the circulatory system. It's even said he performed surgery to remove cataracts, which isn't thought to have been repeated for another 1,000 years. So, it shouldn't surprise us that Galen also worked tirelessly to care for the infected during the pandemic, treating hundreds of victims. His writings are what provide the most information about that ancient plague. He described its symptoms as producing 'a full-body, black, and pustular rash that eventually scabbed over and fell away.' In any event, although few of his works remain, Galen's influence upon the field of medicine would be unmatched until the Arab physician Abu Bakr Muhammad ibn Zakariya al-

Razi. However, the other person of the 2nd century CE to make a notable contribution to science was also the next contributor in the development of the theory of relativity: Claudius Ptolemy.

Not much is known of Ptolemy's birth, early life, or death. The first confirmed date mentioning him is in March of 127 CE, while the last date (the last date that can be verified) is in February of 141 CE. Yet, while we lack a detailed biography of Ptolemy's life, we are able to make a few assumptions based on his name. Claudius Ptolemy is the combination of the Greek-Egyptian name *Ptolemy* and the Roman *Claudius*.

After the death of Alexander the Great and the fracture of his empire, Ptolemy I Soter became Pharoah in Egypt. This former general of Alexander established the approximate 300-year reign of the Ptolemaic dynasty. Therefore, the surname Ptolemy suggests a Greek lineage native to Egypt. On the other hand, his given name – Claudius – was a very common Roman name, indicating that he probably had Roman citizenship. Still, it must be acknowledged that most of what we know about Claudius Ptolemy's biographical history can be, at best, only inferred. But this is not the case when it comes to his scientific achievements.

Ptolemy was a polymath; his scientific achievements can be found in fields as varied as cartography, geography, and trigonometry. For instance, his is the earliest known, still surveying table of trigonometric functions. He made detailed maps of the ancient world, listing longitude and latitude for nearly 8,000 cities, some of which would be unknown today if not preserved in his work. He also did extensive research in the field of optics. And yet, he is not primarily remembered for any of these. Instead, Ptolemy's most exceptional contribution to science was that he did what no astronomer after Aristotle could. Ptolemy succeeded in *patching* the principal flaw in Aristotle's theory of motion. And in so doing, he also introduced the first elements of relativity into the branch of physics that we now call *mechanics*.

Yes, to solve the foremost problem with Aristotle's laws of motion (i.e., to understand the movement of the planets), the concept of relative motion had to be introduced. But before we explore Ptolemy's theory, let's take a moment to remind ourselves of what relative motion is; doing so will assist in grasping the nuance of his work. Relative motion is the idea that both an object's position and motion can be viewed differently due to the position and motion of the person(s) viewing the object.

Epicycles and Equants

As discussed, the problem that immediately arose from Aristotle's theory of motion was his description of the movement of celestial objects. Aristotle believed that all celestial objects, being perfect, moved in perfect circular motions around the then-supposed center of the universe, the Earth. And sure enough, the stars do trace a nearly perfect circular path across the night sky. However, this is not true of the planets. As viewed from the Earth, the planets move with what we today call *retrograde motion*. Unfortunately, the early Greeks had no way of knowing that the stars they so accurately called *wanderers* were actually planets orbiting the Sun, as was the Earth. And their mistaking the planets for stars only served to add to their confusion. However, to grasp the full extent of the issue, let's examine the retrograde motion of Mars as we now understand.

The Earth is 150 million km from the Sun and completes its orbit every 365 days. Mars is almost 228 million km from the Sun, and its year is almost twice as long – 687 days. As a result, every 24 months, the Earth approaches and then passes Mars. Consequently, as seen from the Earth, as we approach Mars from "*behind*," Mars appears to move in an easterly direction across the night sky; this is how Mars appears to be moving most of the time. But once the Earth has "*caught up*" to Mars (i.e., both planets are aligned in their orbits), Mars, for a time, appears to stand still in the night sky. Then, as the Earth "*passes*" the red planet and begins to leave it behind, Mars, for 70 days, appears to move backward across the sky, east to west – this is the "retrograde" motion. Finally, once the Earth swings around the Sun and begins to approach Mars once more from behind, the red planet appears to move along its typical easterly path. Thus, relative to our vantage point on Earth, we perceive Mars as tracing out a zig-zag pattern across the night sky. (Figure 1-4)

Our appreciation of retrograde motion stems from the fact that we know that the planets orbit the Sun. But most of those in the ancient world believed that the Earth sat at the center of the universe. Therefore, Mars' retrograde motion was not explainable.

It was not until Ptolemy came along and examined the problem that he proposed an innovative solution.

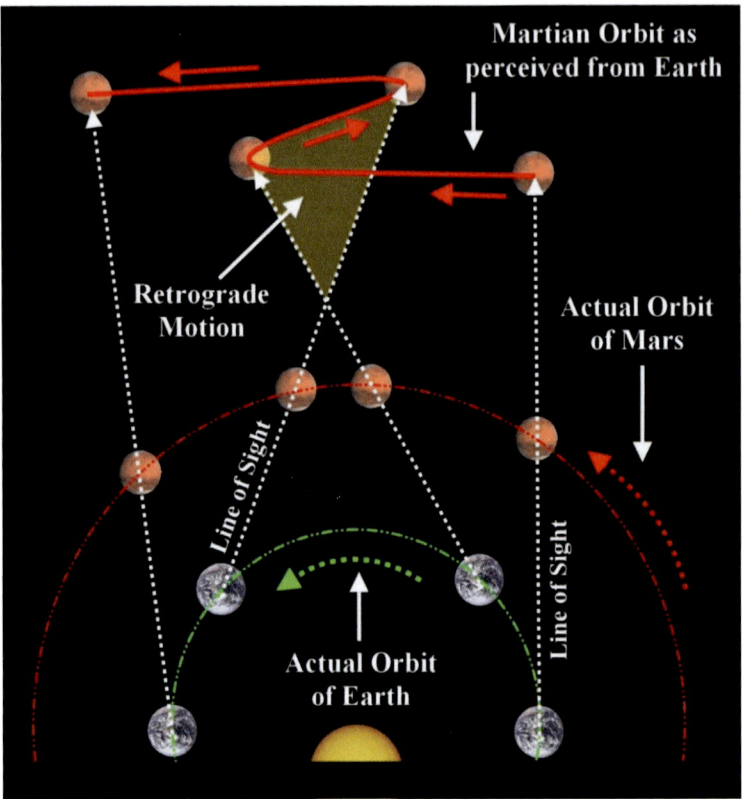

Figure 1-4. The retrograde motion of Mars occurs when the Earth, having a smaller orbit, periodically passes Mars. When this happens, *as seen on a line of sight from Earth* (going right to left), Mars appears to halt its normal progression across the night sky. It then stands motionless for a time before appearing to move backward. Finally, once the Earth has passed the red planet, Mars again travels on its *usual* path.

To resolve the mystery of retrograde motion, Ptolemy suggested that, in addition to circling the Earth in one large circular orbit, the planets also moved in smaller circles. He called these smaller orbits *epicycles*, the center of which was on their primary orbital path. As a result, as viewed from Earth, the retrograde motion of Mars was caused by the planet moving in the opposite direction within its epicycle. (Figure 1-5) What's more, because all the planets exhibited this retrograde motion at regular intervals, Ptolemy was even able to assign sizes and speeds to these epicycles. Evidence that this adjustment to the geocentric model was "correct" was instantly seen. Using Ptolemy's epicycles, astronomers could finally track the progression of a planet more accurately than ever before. Still, even with this startling refinement to the geocentric model and Aristotle's laws of motion, Ptolemy's epicycles alone could not resolve all the issues.

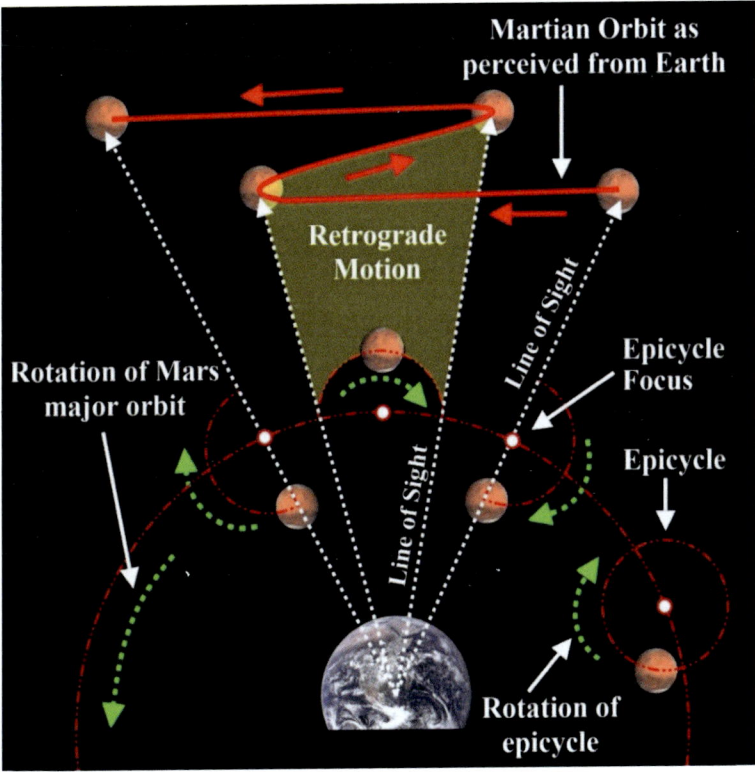

Figure 1-5. Using the concept of epicycles, Ptolemy accounted for the retrograde motion of the planets. And he did so without removing the Earth from the center of the universe.

Another issue with the orbit of Mars was that its velocity also varied depending on <u>when</u> you observed it. This fluctuating speed could not be accounted for, even with Ptolemy's epicycles. But in another spark of genius, Ptolemy also resolved this problem. And in so doing, for the first time in recorded history, the Earth was *successfully* removed from the universe's center. By offseting the Earth from the exact center of the geocentric model, Ptolemy made it possible to account for the variations in speed that the planets exhibited. How so?

Imagine standing alongside a desert road. Next to you is a sign stating that the road's speed limit is 140 km per hour. Off in the distance, you see an approaching vehicle. Now, although you're aware that the car is traveling at the stated speed, it has the appearance of approaching only very slowly because it's so far off. However, something interesting begins to happen as the car gets closer; its speed *appears* to increase. In truth, the car's velocity has remained constant; its speed only has the appearance of increasing. Finally, only when the car passes you does its velocity *appear* to be at a maximum. And then, no sooner does it pass, its speed appears to decrease as it gradually fades into the distance. This example demonstrates a fascinating phenomenon about perspective: An object moving at a constant speed can seem to someone standing far off to be accelerating, even though its speed is unchanged. Aware of this illusion, by moving the Earth slightly off-center in the geocentric model, Ptolemy could account for the relative changes in Martian velocity. (Figure 1-6)

In the geocentric model, the offset Earth is called an *equant*. The result was, as Mars approaches the Earth (just as with the example of the car), its speed *appears* to increase gradually. Then, once Mars passes the Earth at what looks to be its maximum velocity, the planet seems to decrease in speed as it moves away. Thus, by introducing epicycles and moving the Earth slightly off-center, Ptolemy seemingly resolved *most* of the geocentric model's problems. But Ptolemy still wasn't finished.

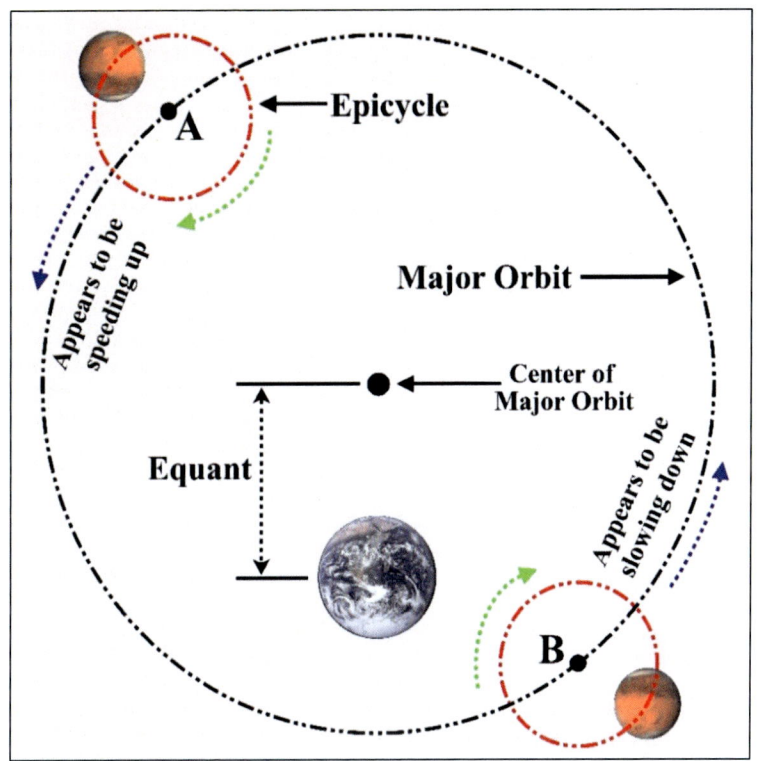

Figure 1-6. By offsetting the Earth in the geocentric model (called an equant), as Mars approaches the Earth, its velocity will *appear* to increase. Then, once it has passed, its velocity will *appear* to decrease.

Planetary Order

A disagreement that persisted among ancient Greek thinkers, even prior to Aristotle, was the exact order of the planets in the geocentric model. The location of particular objects in the night sky was generally agreed upon; for instance, everyone agreed that the object nearest to Earth was the Moon. But that conclusion was straightforward since the Moon's surface features are discernible to the naked eye, which meant that it was closer than the other celestial objects. Besides, Greek mathematicians (using geometry) had already made rough estimates of the Moon's distance from the Earth.[*] Yet, to use geometry to determine the distances between the Earth and the other planets was impossible without a telescope, which wouldn't appear for another millennium. However, there were other methods available that could provide the answer.

The order of the outer planets was deduced based on their *sidereal* years.[†] For example, Saturn has a sidereal period of 30 years. Therefore, Saturn was correctly determined to be farthest away from the Earth of the then-known planets. Jupiter has a sidereal period of 12 years. And Mars has a sidereal period of 2 years. However, this method didn't work for the inner planets. Thus, the geocentric model could not say where Mercury, Venus, and the Sun orbited the Earth. But Ptolemy was able to resolve the issue using his epicycles. He placed Mercury in orbit of the Earth after the Moon, then Venus, and finally, fourth in orbital distance from the Earth, was the Sun. How did Ptolemy reach this conclusion?

---

[*] The first person known to have attempted to calculate the distance between the Earth and the Moon was Aristarchus of Samos (310–230 BCE). Aristarchus used simple geometry and the relationship between the Moon and the Earth's shadow during a lunar eclipse to make his calculations. The second person to attempt to calculate the distance was Hipparchus of Nicaea (190-120 BCE). Hipparchus used the parallax principle, which is the apparent change of position of an object based on where the observer views it. It's like viewing the same car accident but on different sides of the street. Hipparchus provided the most accurate calculation of the Moon's distance, stating that it was some 361 600 to 425 200 km away. The actual distance between the Moon and the Earth is 384 400 km.

[†] A solar year is measured from spring equinox to spring equinox. But the length of a year can also be determined in the night sky. For example, the Earth is on the opposite side of the Sun every six months. Therefore, some of the constellations visible in the night sky also change every six months; this alternating of visible constellations is called a sidereal year. So, in the winter, the constellations Orion and Gemini can be seen in the night sky. In the summer, the latter disappears, and the constellations of Cygnus and Sagittarius are visible. Consequently, as ancient astronomers followed the planets, they found that each planet took a specific number of years before it passed in front of a particular constellation again. This period was the planet's sidereal year, which corresponds precisely to its solar year (the time it takes for the planet to orbit the Sun).

One of the curiosities regarding Mercury and Venus was that they *always* appeared in conjunction with the Sun; they are always visible in the night sky either just before sunrise or right after sunset. In contrast, the other planets appeared throughout the night sky and at different times throughout the night. This behavior is easily explained using the heliocentric, or Sun-centered, model. Mercury and Venus orbit closer to the Sun. So, as we look for them in the night sky, we must always look *toward* the center of the solar system (i.e., nearer the Sun). But the outer planets have larger orbits than the Earth. Thus, to find these in the night sky, we typically look away from the center of the solar system. However, the geocentric model could not explain why Mercury and Venus always appeared in conjunction with the Sun. Using epicycles, Ptolemy could.

Ptolemy suggested that, after the Moon, the planets proceeded in the following order: Mercury, Venus, the Sun, Mars, Jupiter, and finally, Saturn. His solution to why Mercury and Venus always appeared in conjunction with the Sun was to align their primary orbits with the Sun perfectly. In doing so, these inner planets would never be seen in the night sky opposite the Sun. Next, he suggested that the diameters of the inner planets' epicycles were just slightly larger than the Sun's diameter. Thus, one would only see Mercury and Venus when their respective epicycles brought them to either side of the Sun – either right before sunrise or right after sunset. (Figure 1-7) Otherwise, when their epicycles had them in front of the Sun, the Sun's light obscured them. These solutions were extraordinary intellectual leaps. Indeed, Ptolemy so drastically altered the geocentric model that, from then on, it was synonymously called the Ptolemaic model. However, this didn't mean that Ptolemy's new model lacked flaws.

Ptolemy's model worked well in predicting the positions of the planets over the short term. However, as one attempted to track the planets' positions over an extended period, his system began to break down. Frankly, it was just too complicated. And as the centuries advanced toward the introduction of the telescope, so did the techniques for tracking the planets. Soon, the only way to prop up Ptolemy's system of epicycles and equants was to add more of them! Thus, by the 16th century, some outer planets

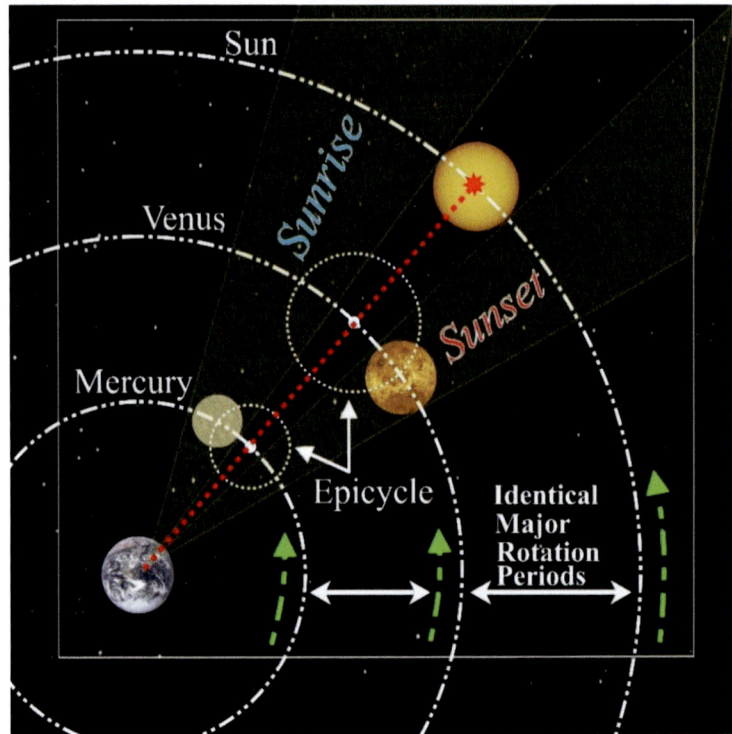

Figure 1-7. Ptolemy suggested that both Mercury and Venus orbited the Earth in perfect alignment with the Sun (dashed red line). He also stated that their epicycles were just a bit larger than the diameter of the Sun. The result was that Mercury and Venus could only ever be seen before sunrise or after sunset.

required as many as seven different sized epicycles to accurately describe their orbits using the Ptolemaic system. Still, even ignoring this problem, the Ptolemaic system's main dilemma arose, not from the orbits of the planets, but rather from the orbit of the Moon. Although it may not appear so, the Moon's orbit is exceptionally complex.

Like the other planetary bodies, the Moon follows an elliptical orbit. Therefore, its distance from the Earth varies by up to 42 000 km. (Figure 1-8) And yet, despite this seemingly large distance, the Moon's size differs by no more than 14 percent throughout the month. (Figure 1-9 [*]) However, to keep his mathematically model consistent, Ptolemy needed to give the Moon an epicycle. And once he did this, he then was forced to accept the consequences.

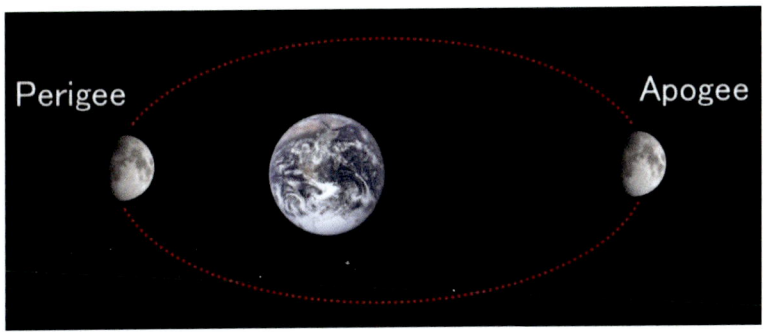

Figure 1-8. The Moon's orbit about the Earth is elliptical, not perfectly circular. Therefore, its distance from the Earth varies by as much as 42 000 km. As a result, the Moon's size appears to change.

Figure 1-9. The Moon is at its perigee when it is closest to the Earth in its orbit. At that point, its diameter changes by no more than 14% from when it is at its apogee. This is far less than Ptolemy's suggested "doubling" in size.

---

[*] Image credit: NASA

An epicyclical Moon would need to be at its maximum size in the night sky when in its quadrature.* Thus, Ptolemy was forced to assume that this was when the Moon was at its perigee.† As a result, his calculations indicated the Moon should appear two times larger in its quadrature than at its syzygy‡ (the position he believed to be the Moon's apogee). However, given the relative nearness of the Moon to the Earth, it could easily be seen that it did not have such a large epicyclical motion. (Figure 1-9) Therefore, it was also clear that Ptolemy's estimates were grossly exaggerated. Making things even more difficult for Ptolemy is that the Earth also tilts on its axis. As a result, the Moon's position in the sky also varies north to south by 5 degrees annually. Reconciling these motions using epicycles and equants alone was impossible. So, as was the case with the geocentric model that preceded it, the Ptolemaic model also had flaws and needed improvement.

Nevertheless, the problems with Ptolemy's system do not diminish what he accomplished. Most importantly, he introduced the world to the idea of relative motion – that the movement of the planets, as seen from the Earth, was relative. His suggestion of fixing the geocentric model with epicyclical motion, where smaller circular orbits traced out a path along a larger orbit in the geocentric model, was beyond innovative. He was also the first to *convince* the general populace that the Earth did not sit at the *exact center* of the universe. These were all necessary steps forward in the development of the theory of relativity. In the final analysis, the Ptolemaic system worked sufficiently well to last for 1,400 years, a testament to his ingenuity.

---

* The quadrature occurs when the Moon is at a right angle to the Earth, as viewed from the Sun; this occurs during its first and third quarters when the Moon appears half full. Ptolemy believed that this was when the Moon was closest to the Earth.

† The perigee is when a satellite's orbit (e.g., the Moon) brings it nearest Earth. The apogee is the point of a satellite's orbit when it is farthest from the Earth.

‡ A syzygy is when three celestial objects align, most often applied to the Sun, Earth, and Moon's alignment. The Moon's syzygy occurs twice in its orbit. Once during a new moon (when the Sun, Moon, and Earth are aligned, in that order). Then once during a full moon (when the Sun, Earth, and Moon are aligned, in that order).

Chapter 2

With the conclusion of Marcus Aurelius' reign, the 207-year *Pax Romana*, or Roman Peace, came to an unsuspecting end. From 180 CE onward, according to most historians, the Roman Empire began its arduous decline. Soon after, they endured a civil war (193-197 CE) that squandered the empire's resources and diverted the army's attention away from its borders and the invading Germanic tribes. In addition to constant political intrigue, persistent economic downturns soon made it impossible for one man to manage such a large empire. In the end, the strain of controlling so vast a territory became too much. Hence, in 285 CE, Emperor Diocletian split the empire into two separate states. And instantly, the Western Roman Empire came into existence, with its capital in Rome and the Eastern Roman Empire in Byzantium. In hindsight, this final act made the Western Roman empire's collapse in 476 CE seem all but fated.

With the fall of the Western empire, the authority of Roman governors quickly declined in Europe, and local peoples returned to their tribal heritage. In time, this eventually gave rise to a feudal system headed by royal families and vassals who ruled over lowly and disadvantaged peasants. Then came the numerous conflicts and religious crusades. Europe had entered its *Dark Age*, which lasted for centuries – from 476 CE to 1300. It was a time in western civilization when cultural and social advancements came to a near standstill. And though it cannot be said that there was no scientific progress during this period in Europe, there were no significant contributions to the theory of relativity. However, as with all things, this social stagnation of the Dark Ages eventually began to yield to another age, an era that signaled a rebirth of Western civilization. It started at the close of the 13th century with the arrival of what would become known as the Renaissance.

As city-states struggled to feed their fast-growing population, the European continent wrestled with increased mismanagement of its farmlands. At the same time, ongoing conflicts raged throughout 13th century Europe. Accordingly, the situation reached a critical juncture by the middle of the 14th century. The circumstances were ideal for the disaster that inevitably follows such events: famine and disease.

The *Black Death* arrived in Europe in October 1347. It's estimated to have killed some 75 to 200 million people – a figure that may well have been 30 to 60 percent of the continent's population. But even more significantly, most of those who died were lower-class citizens. But it was this tremendous shift in the population that brought extraordinary social changes to Western Europe. More specifically, the general vocational skills that had been commonplace among underprivileged peasants prior to the plague became sought-after trades. And it was this shift in demographics that led to the emergence of a wealthy merchant class, one entirely disconnected from any established aristocracies.

The term *Renaissance* comes from 19th-century French; it means *rebirth*. It was a rebirth in Western Europe of Greek and Roman culture – art, literature, and architecture. And while interest in these subjects existed before the 14th century, it was only found among the nobility or monks serving in monasteries. But that all changed once this new, wealthy merchant class appeared. And when coupled with Johannes Gutenberg's printing press, once common European citizens now began to experience the culture and thoughts of the ancient Greeks and Romans. This rebirth began sometime before 1350 and lasted some 250 years. Its origin was in Italy but, by the late-15th century, it had reached the Polish-Lithuanian Commonwealth, which was the birthplace of the next significant contributor to the theory of relativity.

Nicolaus Copernicus (1473-1543)

As can be surmised from their surname *Copernicus*, the paternal side of his family appears to have originally been copper

merchants. Copernicus' father, also named Nicolaus, was born in Kraków, though it's unknown in what year. But, in 1458, he moved to Toruń to enlarge his political and business ambitions. By this point in European history, the Thirteen Years' War between Poland and the Teutonic Knights had been raging for two years. Then serving as a city magistrate, Nicolaus Sr. had chosen to support the Prussian Alliance.

Copernicus' mother, Barbara Watzenrode, was born in Toruń, Poland, sometime around 1440. She was the daughter of Lucas Watzenrode the Elder, a wealthy landowner, and judge allied with the Prussian Alliance. At length, in 1462, the same year her father died due to injuries sustained in the war, Nicolaus and Barbara were married. Eleven years later, their last child Nicolaus was born. Nicolaus Copernicus (Jr.) was the youngest of four children. His eldest brother was Andreas, and his two older sisters were named Barbara and Catharina.

The young Nicolaus' early childhood was spent in Toruń, where the family held much property and had quite a few servants. And since this region was under dispute, it was often traded back and forth between Germany and Poland. So it's fair to assume that, growing up, Copernicus likely learned to speak both German and Polish. Additionally, given his family's affiliation with the church and their affluent societal status, he would have also been taught to speak Latin. However, like Aristotle, Copernicus and his siblings were beset by tragedy early on when their father died sometime after 1483. It thus fell to their uncle, Lucas Watzenrode the Younger, to take on the role of the family patriarch.[*]

By the time Copernicus and his brother were old enough to enter university, their uncle had risen to Prince-Bishop of Warmia, a region in the Polish-Lithuanian Commonwealth. Wishing to bolster his position within the church by surrounding himself with trusted allies, the Prince-Bishop encouraged Copernicus and his brother to pursue careers in theology. Both men did so upon enrolling at the University of Kraków in 1491. And it was while at university that, aside from studying Latin and philosophy, Copernicus was introduced to astronomy. However, the latter course was not exactly what we today consider the *science* of astronomy.

University astronomy courses of the 15th century contain two incredibly divergent aspects as part of their curriculum. First was the mathematical aspect, which included the works of Aristotle and Ptolemy. Knowing astronomical math was necessary to

---

[*] Copernicus' mother is not believed to have remarried; she died sometime between 1495 and 1507.

understand the era's calendars; it also taught one to navigate by observing the stars. However, the second aspect of Renaissance astronomy was the study of astrology and horoscopes, which was heavily practiced within the church.[*]

Copernicus completed his studies at the University of Kraków after about four years, though without earning a degree. Then, in October 1496, he enrolled at the University of Bologna to study canon (i.e., church) law. As fortune would have it, this was where he met Domenico Maria de Novara. De Novara served as professor of astronomy at the University of Bologna from 1483 to 1504. Like Copernicus, astronomy was not de Novara's primary field of study; his doctorate degrees were in art and medicine. Still, it was their shared love of astronomy that fostered their close friendship and led to their collaboration on several astronomical projects. The most important of these collaborations occurred just months after Copernicus enrolled at the school. On March 9, 1497, the Moon eclipsed Aldebaran, a celestial phenomenon that would change Copernicus' life.

Copernicus graduated from the University of Bologna in 1500 with a doctorate in medicine and soon became his uncle's royal physician and advisor. However, after viewing the Moon's eclipse of Aldebaran, his passion forever remained astronomy. Indeed, the theory of the universe that he would eventually propose became known as the *Copernican model*; and the scientific shift, the *Copernican revolution*. However, Copernicus would receive no notoriety for his discoveries while alive. On the contrary, Copernicus did everything possible to prevent his theory from being publicized.

Officially, the 15th-century Catholic Church endorsed the Ptolemaic model, and to dispute that view was considered heresy. And given that Copernicus studied church law, he was undoubtedly aware of what could happen if he defied church dogma. So, although convinced that the geocentric model was wrong, Copernicus never publicly criticized it. Instead, while alive, he only allowed a few trusted friends to read a hand-written

---

[*] Prominent Popes – such as Julius II (1503 - 1513), Leo X (1513 - 1521), and Paul III (1534 – 1549) – were all known to regularly consult astrologers.

treaty of his heliocentric (Sun-centered) model. It wasn't until after his death that his treatise *De revolutionibus orbium coelestium* (On the Revolutions of Heavenly Spheres) was published. (Figure 2-1) It even included a 3-plus page apology to Pope Paul III.[*] In any event, within his treatise, Copernicus made one of the most important breakthroughs in human *thought*. Because the logical proof that the Earth was not the center of the universe was important, but not just scientifically; it was also psychologically significant.

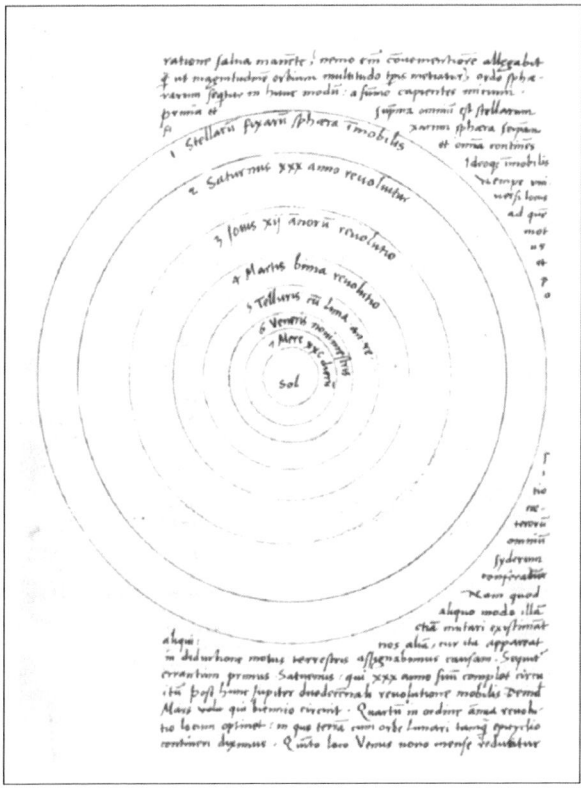

Figure 2-1. One of Copernicus' original manuscripts, now located at the Jagiellonian Library in Kraków, Poland.

---

[*] Death was also no guarantee against posthumous church judgment. Condemned heretics in the 15th century were routinely exhumed and bodies defiled for perceived sins they had committed against the church when they were alive.

Prior to 1543, humanity insisted on imagining that the Earth resided at the center of the universe. And this mistaken idea was used to support the notion that, within the universe, humans were central in status. However, when Copernicus introduced the heliocentric model, he proposed a different notion. As a result, as people accepted that the Earth was not the center of the universe, they were ushered out of an egocentric frame of reference. The universe operated on a far grander scale than they could have ever imagined, one that did not center around *us*. And this fantastic mental insight began, as mentioned, with the Moon's eclipsing of Aldebaran in 1497.

Aldebaran Occultation

For Copernicus, the most convincing evidence that the geocentric model was wrong occurred during the occultation of Aldebaran. An occultation occurs when one celestial object passes in front of another, hiding the second object from view, such as when the Moon blocks the Sun during a solar eclipse. Aldebaran is a red giant located in the constellation of Taurus. On March 9, 1497, the Moon passed in front of this very bright star. However, what was unique about this occultation was that the Moon was in its quadrature or half-moon. As a result, Copernicus could watch as Aldebaran disappeared behind the Moon's dark side and reappeared adjacent to its light side. Knowing the orbital period of the Moon and the time that Aldebaran was obscured, Copernicus could use geometry to calculate the Moon's parallax.* Why was

---

* Parallax is the apparent change of position of an object due to its being viewed from differing locations. The best example of this occurs when there is a solar eclipse. When the Moon passes in front of the Sun, the eclipse's appearance depends on where on the Earth you are located or, more accurately, your latitude. To see a full solar eclipse, you must be located directly beneath the Moon's shadow. The farther north or south you are from this exact path of the Moon's shadow, the more the full eclipse becomes a partial eclipse. For example, a full eclipse in London would be a partial eclipse 97 km away in Cambridge. And 322 km away in Manchester, an eclipse would not be seen at all. Everyone could watch the Moon move across the sky in each of these locations. But, depending on where you are, the Moon appears to be in vastly different positions relative to the Sun. However, knowing the distance between these various cities and the Moon's angle in the sky, using only geometry, you could accurately calculate the distance between the Earth and the Moon.

this significant? Because it could now be proven that the Moon's distance during its syzygy and quadrature was not as great as Ptolemy supposed. Or, in other words, *the Moon had no epicycle.* And given that epicycles were the cornerstone of the Ptolemaic model, Copernicus and de Novara were now convinced that the Ptolemaic and geocentric models of the universe were wrong. But how did they know to replace it with the heliocentric model?

Heliocentric Model

It doesn't appear that Copernicus had the goal of disproving the geocentric model. Instead, it seems that he merely wanted to show that the heliocentric model was, by comparison, a simpler alternative. In truth, it wasn't even possible in Copernicus' day to prove or disprove that the Earth revolved around the Sun; that credit would go to the English astronomer James Bradley (see Chapter 3). And while we know that de Novara didn't believe in the Ptolemaic system, we don't know if replacing it with the heliocentric model was his suggestion. Therefore, we can't say how Copernicus settled on the heliocentric model, though there is inconclusive proof that he may have gotten the idea from a secondary source. In his book, Copernicus quoted the writings of one Philolaus, the Pythagorean – an ancient Greek who lived before Aristotle.

[Earth] moves around the fire with an obliquely circular motion...

Philolaus the Pythagorean –
as quoted by Nicolaus Copernicus
On the Revolutions of Heavenly Spheres

To be sure, the idea that the Earth revolved around the Sun was not a new concept; it had first been suggested in the 5th century BCE. And given the admiration for all things Greek during the Renaissance, it's not improbable that Copernicus came across Philolaus' writings while at university. However, others present an equally compelling argument against his getting the idea from the Greeks.

Aware of the Catholic Church's high regard for ancient Greek philosophy, some say Copernicus found Philolaus' quote only *after* he had conceived of the heliocentric model. Thus, his quoting Philolaus was merely a tactic to preemptively protect himself from the Inquisition. Indeed, given that the quote is found in his apology to the Pope, the above argument also has merit.

So, how Copernicus came up with the heliocentric model remains a mystery. But what's important is that the more he studied the heliocentric system, the more he realized that it could resolve all the significant issues associated with the Ptolemaic model. For example, it solved the Moon's lack of an epicycle and the dilemma as to why Mercury and Venus appeared in the sky only near sunrise and sunset. Additionally, it made the retrograde motion of Mars and the other planets easily explainable. Yet, what made Copernicus even more convinced of the heliocentric model was another discovery made just five years before he saw the occultation of Aldebaran.

A major discovery still being spoken of in 1497 was Christopher Columbus' confirmation that the Earth was spherical. Consider some of the consequences that this discovery must have brought to Copernicus' mind. Like the heliocentric model, the idea that the Earth was a sphere was also not new; it too was an idea that most had simply refused to accept. And if all the then-known planets were spherical, and now the Earth was proven to be a sphere, there was only one conclusion remaining. The Earth was just another planet orbiting the Sun. Thus, besides the occultation of Aldebaran, confirmation of a round Earth was only further proof of the heliocentric model. Yet, Copernicus would make another critical contribution to the theory of relativity.

Now convinced that the heliocentric model was correct, Copernicus began to explore its consequences. And in doing so, he connected the heliocentric model of the universe with relative motion and gravity.

Accordingly it was necessary for there to be less water than land, so as not to have the whole earth soaked with water – since both of them tend toward the same centre on account of their weight…

And from all that I think it is manifest that the land and the water rest upon one centre of gravity...

Now that it has been shown that the earth is a globe, I think we must see whether or not a movement follows upon its form and what the place of the earth is in the universe. For without doing that it will not be possible to find a sure reason for the movements appearing in the heavens... For every apparent change in place occurs on account of the movement either of the thing seen or of the spectator, or on account of the necessary unequal movement of both. For no movement is perceptible relatively to things moved equally in the same directions – I mean relatively to the thing seen and the spectator.

Nicolaus Copernicus –

On the Revolutions of Heavenly Spheres

As we read Copernicus' statements above, many of the core concepts of relativity are present. First, he rightly concluded that since the Earth was a globe, all objects must fall, not toward their similar elements, but instead toward the center of the Earth. (Figure 2-2) He correctly deduced that all objects fell toward the

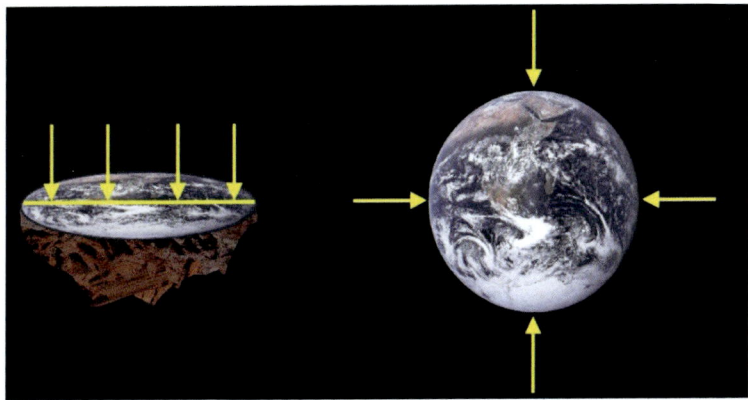

Figure 2-2. According to Aristotle, in the flat-earth model, all objects move toward the elements they are composed. However, after Columbus confirmed that the Earth was spherical, Copernicus correctly realized that all objects must fall toward Earth's center.

Earth due to their weight (or mass). Lastly, for the first time in history, Copernicus made an explicit statement regarding the notion of relativity; that all motion was relative to the one observing the motion (*"the spectator"*) and the object that was in motion (*"the thing seen"*). Lastly, he added, *"or on account of the necessary unequal movement of both."* So, Ptolemy's epicycles and equant demonstrated that he understood how to *apply* relative motion to "fix" the geocentric model of the universe. But Copernicus unmistakably stated the concept of relativity, and not just with regard to planetary motion. Copernicus had made a fantastic mental leap forward. However, as with the Ptolemaic system, Copernicus' system also had a singular flaw.

Although Copernicus had abandoned the ancient belief in a geocentric universe, he still held onto other outdated notions. For example, he tried to fit into his heliocentric model the Greek idea that celestial objects orbited in perfect circles, something the math could not support. As a result, although his model resolved many of the issues associated with Ptolemy's system, the Copernican model proved to be just as inaccurate in tracking the planets. It would take another giant in the field of astronomy, Johannes Kepler (1571-1630), to add the final correction to the heliocentric model.

Johannes Kepler (1571–1630)

Kepler improved the Copernican system by mathematically proving that the orbits of the planets <u>could not</u> be perfectly circular; they were instead elliptical. This slight change finally made the motion of the planets mathematically predictable. Then, in 1619, Kepler formally proposed his now-famous 3-Laws of planetary motion:

1. The planets revolve around the Sun in elliptical orbits, with the Sun at one focus.

2. An imaginary line connecting the Sun to a planet will always sweep equal areas in equal times.

3. The square of any planet's orbital time (T) is proportional to the cube of its average orbital radius (*R*). Expressed mathematically as:

$$T^2/R^3 = \text{constant}$$

When Kepler's laws of motion were added to the Copernican system, it now became possible to predict the positions of any of the then-known planets *years* in advance.[*] Still, even with remarkably accurate astronomical calendars as proof, it didn't convince others of his day to let go of the Ptolemaic system.

It was a strange contradiction: Copernicus' book, *On the Revolutions of Heavenly Spheres*, was widely published, more than 500 first- and second-edition copies remain today. Yet, very few of his contemporaries were willing to concede that the Earth revolved around the Sun. For most, the heliocentric model was a counter-intuitive notion that went contrary to everyday experience. It proved to be too fantastic of a mental leap. And yet, the biggest issue with accepting the heliocentric model was probably the justified fear of religious retribution.

There was a legitimate fear of the church's response to the heliocentric model; for this reason, Copernicus did not allow his treatise to be published until after his death. In contrast, Kepler, who lived about 30 years after Copernicus, openly advocated the heliocentric model and did suffer religious resistance. However, Kepler was spared the dreaded Inquisition because the protestant reformation partially shielded him in Germany.

Additionally, Copernicus presented his book more like a mathematical endeavor than suggesting that the Earth revolved around the Sun. For this reason, it didn't immediately appear to

---

[*] A planet can have a circular orbit; it's just improbable that it would remain that way. A planet with a perfectly circular orbit would continue in that orbit so long as no other forces were acting on it. For instance, the Earth lacks a perfectly circular orbit because of the large gas giants. Thus, each time Jupiter and the Earth are on the same side of the solar system, Jupiter's gravity tugs on the Earth, not enough to pull it away from the Sun, but just sufficient to perturb its orbit. The result is that the Earth's orbit is elliptical. Ultimately, Copernicus' heliocentric model, though conceptually correct, required mathematical refinement.

challenge church dogma. This latter tactic was effective – a tactic that should have been copied by the next contributor to the theory of relativity. Because this next scientist's unyielding personality eventually forced him to endure the full impact of church censure. And while he was spared the death penalty based solely on his popularity, he was, nonetheless, forced to spend the remainder of his life under house arrest.

Galileo Galilei (1564-1642)

As was true with all the contributors we've considered so far, the life and achievements of Galileo Galilei are certainly worth further study. For example, in addition to his contributions to science and math, Galileo was a pioneering inventor. Many of the devices he created, such as his early thermometer, were the forerunner of laboratory devices still in use today. It is, therefore, for good reason that Galileo is considered the *father of modern science*. However, for the sake of brevity, we will begin by discussing a few episodes in his early life, starting with some facts about his father. Vincenzo Galilei was born in Florence, Italy, in 1520. And while he's primarily remembered as Galileo's father, Vincenzo's accomplishments are also worthy of note.

Vincenzo was an accomplished lutist, so much so that, at the peak of his career, he had gained several wealthy and influential patrons. He was the 16th-century version of a rock star. Then, at the age of 43, Vincenzo again became an apprentice to Gioseffo Zarlino, one of Renaissance music's most prominent composers. In time, Vincenzo's musical theories laid the foundation for the development of baroque, the forerunner of modern classical music. Vincenzo also made advances in the study of acoustics; through experiments with vibrating strings, he discovered that when one string was half the length of another, the two will resonate when played simultaneously. This interval is what we now call an octave. So, it appears that there was a natural talent for music and mathematics within the Galilei family.

In contrast, not much is known of Galileo's mother, Giulia. We know her maiden name was Ammannati. She was born in Pescia, Italy, in 1538, which made her almost 20 years younger than Vincenzo. It's also known that her father was a wealthy lumber merchant. However, records show that her brother Leone paid the dowry. So, her father had probably died when she and Vincenzo married in 1563. Finally, after getting married, Vincenzo and Guilia made their home near Pisa, Italy, where Galileo was born on February 15, 1564; he was the eldest of seven children.

Vincenzo was determined that his son receive a good education. So when the family moved to Florence in 1572, he left Galileo in Pisa for two years with Muzio Tedaldi, a relative on his mother's side. Then, in 1574, after Galileo had rejoined his parents, he arranged for his son to be tutored by a man named Jacopo Borghini. However, a 17th-century biographer states that Galileo didn't think too highly of him. Three years later, Galileo was sent to study at the Camaldolese Monastery at Vallombrosa, roughly 38 km east of Florence. As he engaged in religious studies, it was here that Galileo was most likely taught Latin and Greek. From all accounts, Galileo enjoyed his time there; he even considered becoming a monk. Nevertheless, perhaps when around 17, Galileo's father removed his son from the monastery and returned him to Florence. Vincenzo had resolved that Galileo would pursue the more lucrative career of medicine.

Thus, by 1581, although he was not particularly interested in the subject, Galileo was studying medicine at the University of Pisa. But, as chance would have it, while at the university, he was introduced to mathematics by Filippo Fantoni. At long last, outside of monastic life, Galileo had found a field of study that he was both skilled at and enjoyed. Just as important, his mathematical talent was evident to others, even as a first-year student. Galileo was soon introduced to the Tuscan Court mathematician Ostilio Ricci. Immediately impressed with the young man, Ricci took it upon himself to try and convince Galileo's father to allow his son to take courses in mathematics, in addition to medicine. Only reluctantly was Vincenzo persuaded.

However, as Galileo grew more proficient in mathematics and his reputation increased, he decided to abandon the study of medicine without his father's permission. Indeed, by 1585, even without earning a degree, he decided to leave the University of Pisa. Galileo had become so well-known that he had determined to find a position as a mathematics instructor. And as odd as this decision might have appeared, it was a decision that put him on the path to physics' stardom.

By the summer of 1586, Galileo had published his first book, *La Balancitta* (The Little Balance). It described Archimedes' method of finding the relative densities of substances using a technique known as hydrostatic balancing. By this time, Galileo had become so well-known that he even received an invitation from the prestigious Florence Academy to lecture on, of all things, the dimensions and location of hell in Dante's Inferno. A year later, he returned to the University of Pisa, where he was appointed a mathematics professor. Soon after, he was offered a position at the University of Siena. All seemed to be going well for Galileo until, in 1591, his father died. Now, as the eldest son, he became responsible for caring for his immediate family, a frightening task given that he was barely earning enough to support himself. Yet, as chance would have it, Galileo's financial situation would quickly change.

By the 1590s, Galileo's talents became known to the mathematician and philosopher Guidobaldo del Monte. As a result, the latter recommended that Galileo be made Professor of mathematics at the University of Padua, located about 290 km northeast of Siena. This university was one of Europe's earliest, most renowned, and still existing schools. And since he would receive a threefold increase in pay, Galileo readily accepted the position and remained there for the next 18 years. Most notably, it was there that Galileo would make his first public statements against Aristotelian theories. His objections would be based on what would later become known as Kepler's Supernova.

As early as 1598, in letters written to Kepler, Galileo had admitted to being "a Copernican." Still, he remained cautious about making public statements against Aristotle's teachings. But he became convinced of his opinion in 1604 when a supernova occurred in the Milky Way, just 20,000 light-years from the Earth. It was one of the few in human history to be close enough to view with the naked eye. What's more, the light from this exploding star was so intense that it could be seen during the day. And although Kepler was not the first to observe it, he was the one who successfully tracked the supernova's progress for more than a year, beginning on October 17. Galileo attempted to measure the distance to the supernova using parallax but could only determine that it was extremely distant. However, more important than ascertaining its distance, Galileo realized that this exploding star directly contradicted Aristotle's belief that the stars were unchangeable and eternal. And once convinced that Aristotle was wrong, when Galileo's opinion did become public, his subsequent conflict with the Catholic Church became just as important to history as were his scientific discoveries.

The Telescope

Although many credit Galileo with inventing the telescope, it was the German-Dutch lensmaker, Hans Lipperhey, who did so in September of 1608. However, upon learning of its creation, Galileo enthusiastically applied himself to improving it. Within months, he constructed a telescope with a magnification of 8X, which allowed him to see just how extensive the Moon's scars and craters were – more proof that celestial objects were as *perfect* as Aristotle had theorized. But Galileo's most profound discovery using the telescope occurred about two years after it was invented.

In 1610, Galileo pointed his newly constructed 20X telescope toward Jupiter. For eight weeks, he tracked the positions of four never-before-seen objects that appeared to be circling that planet. As an experienced scientist, he immediately recognized that they were moons. (Figure 2-3) It took Galileo another two years to calculate the orbital periods of Jupiter's moons. Ironically, the reason it took him so long arose from the fact that he, an admitted Copernican, failed to consider the Earth's movement around the Sun in his calculations. Still, the implications were undeniable: The Earth was considered the center of the universe. Therefore, the universe, including our Moon, orbited it. However, discovering moons that had as their center, not the Earth, but Jupiter, it stood to reason that the Earth was not the center of the universe. Galileo's discovery of the Jovian moons was a validation of Copernicus. Still, it would not be until Isaac Newton formulated his laws of motion that the heliocentric model became understandable and generally accepted.

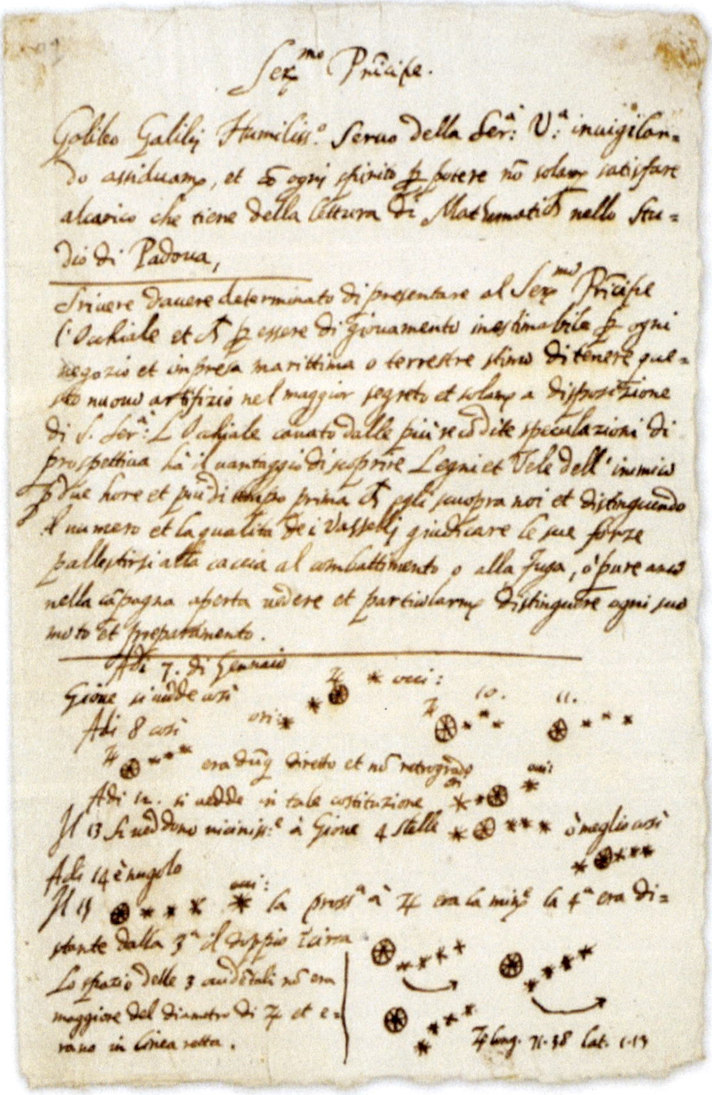

Ser.mo Principe.

Galileo Galilei Humiliss.o Servo della Ser.tà V.a invigilan-do assiduam.te, et con ogni spirito per potere non solam.te satisfare al carico che tiene della Lettura di Matematiche nello Stu-dio di Padova,

Inoltre d'havere determinato di presentare al Ser.mo Principe l'Occhiale et per essere di giovamento inestimabile per ogni negozio et impresa marittima o terrestre stimo di tenere que-sto nuovo artifizio nel maggior segreto et solam.te a disposizione di S. Ser.tà L'Occhiale cavato dalle più recondite speculazioni di prospettiva ha il vantaggio di scoprire Legni et Vele dell'inimico per due hore et più di tempo prima che egli scuopra noi et distinguendo il numero et la qualità dei Vasselli giudicare le sue forze pallestirsi alla caccia al combattimento o alla fuga, o pure anco nella campagna aperta vedere et particolarm.te distinguere ogni suo moto et preparamento.

Adi 7. di Gennaio Giove si vedeva così

Adi 8 così

era dunq. diretto et non retrogrado

Adi 12. si vedde in tale costituzione

Il 13 si veddono vicinissime à Giove 4 stelle o meglio così

Adi 14 è nugolo

Il 15 la prossima à ♃ era la minore la 4a era di-stante dalla 3a il doppio in circa.

Lo spazio delle 3 occidentali non era maggiore del diametro di ♃ et erano in linea retta.

♃ long. 71.38 lat. 1.13

Figure 2-3. These are Galileo's actual notes tracking the orbits of Jupiter's moons. The planet, its moons, and each moon's changing position – found in the bottom, right corner of the document – are clearly discernible in his sketch.

The above discoveries alone would have gained Galileo acclaim. Yet, he still made other significant findings using the telescope. For instance, he was the first to see that the planet Venus experienced phases, which would not happen unless it too orbited the Sun. Galileo discovered that Saturn had rings and that the "milky" haze of the Milky Way was, actually, more stars than could be resolved with the naked eye. And, observing dark sunspots on the Sun's surface, he was even able to verify that not even the Sun was immune to "imperfections." (Figure 2-4) The

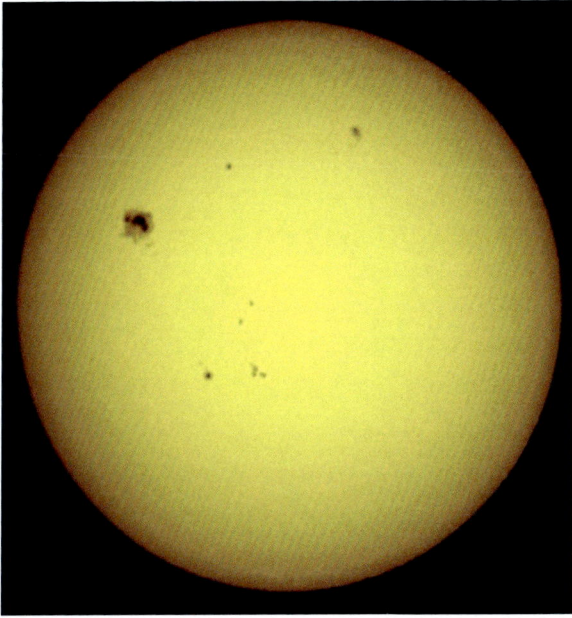

Figure 2-4. The presence of sunspots indicated to Galileo that our home star was not as "perfect" as Aristotle had suggested.

telescope is an excellent example of how science can advance technology that then advances scientific understanding. We can only suppose that if Aristotle had had the advantage of seeing the universe through a telescope, he might well have also reached different conclusions. But, in the end, Galileo had made too many discoveries that could only be explained using the Copernican system. Therefore, he now sought to do what no scientist before him had dared; he sought to refute church dogma – a bold decision that had severe repercussions.

Galilean Relativity

Galileo's theories concerning relative motion are explained in his book *Dialogue Concerning the Two Chief World Systems*. This book was scripted as a debate between two philosophers and a layman. Galileo's goal was to explain why persons standing on the Earth's surface and traveling in orbit around the Sun would not *feel* as if they were in motion.* In addition to this being the first book in history dedicated to explaining relativity, it also refuted the geocentric system. Furthermore, to prove that relative motion was factual, Galileo employed the now-famous analogy of a sea-faring ship.

Within the hold of a ship is a sailor, a small cage containing butterflies, and suspended from the ceiling is a jar slowly dripping water. (Figure 2-5) Now, suppose the ship was motionless, moored at the dock, one could conduct these experiments and get the following results. (1) From a standing position, the sailor could jump almost 2 meters forward; (2) the butterflies, when released, fly effortlessly about the ship's hold; and (3) the dripping water falls directly into a cup below. Galileo next imagined the ship sailing on the open seas, then considered the results of the

---

* Galileo's *Dialogue Concerning the Two Chief World Systems* is considered one of the most important scientific treatises in history. The challenge Galileo faced was that the heliocentric model was labeled blasphemous by the church. Therefore, Galileo had to present his book such that he didn't appear to endorse the heliocentric model officially. The book was presented as a dialogue between 3 individuals: Salviati (who represented Galileo), Sagredo (who represented an undecided layman), and Simplicio (the advocate of the geocentric model). Simplicio, representing the Catholic Church, was named after Simplicius of Cilicia, a 6th-century commentator on Aristotle; however, the name Simplicio had a double meaning. In Italian, it also meant "simple-minded." In the book, the characters discussed the pros and cons of the heliocentric versus the geocentric models. The evidence supporting the heliocentric model was brilliantly presented, making Simplicio look, well, simple-minded. In any event, Galileo concluded the book with Salviati conceding that he was wrong, that the geocentric model was correct. However, Church officials were not fooled. One year after Galileo published his book, he faced the Inquisition, was found guilty and was sentenced to house arrest for the rest of his life. *Dialogue Concerning the Two Chief World Systems* was placed on the church's list of banned books and was not removed until 1835, some 200 years later.

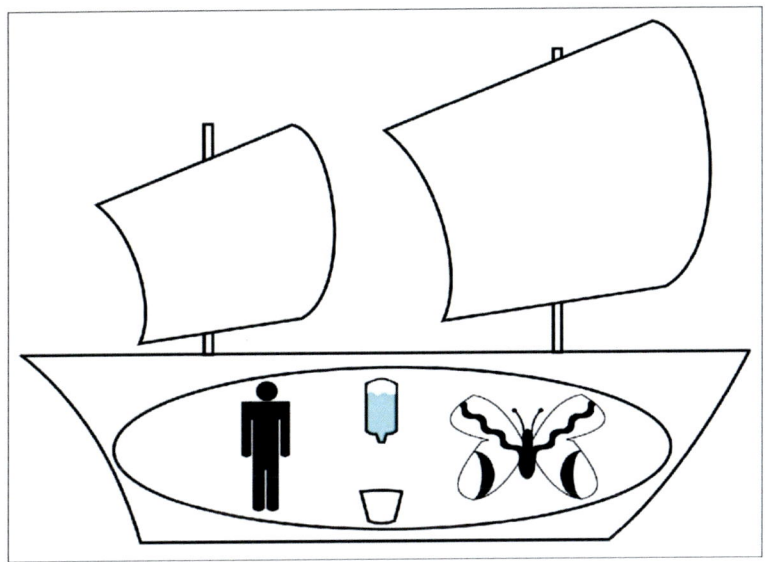

above experiments using Aristotle's logic. Recall, Aristotle thought that objects only moved when being pushed. Let's begin with the sailor jumping.

When the ship was stationary, the sailor could jump forward 2 meters. So now, with the ship in motion at a *uniform* speed, we repeat the experiment. Below deck, the young sailor jumps in the direction of the bow and then back toward the stern. According to Aristotle's laws of motion, while the sailor is in the air, he is not being *pushed* by the ship. Yet, the ship is still being *pushed* forward by the wind. As a consequence, when the sailor lands, measuring the distance of his jump <u>along the ship's deck</u>, what distance will he have covered? The total distance the sailor covers, *according to Aristotle*, <u>should</u> be shorter and greater than 2 meters, respectively. Why?

We know the sailor can cover 2 meters when jumping on a stationary ship. (Figure 2-6a) But on a moving ship, as the sailer jumps toward the bow, the ship's deck is still moving *forward*. Therefore, when he lands back onto the deck, the measured distance would appear to be less than 2 meters. It would be the 2 meters he could jump *minus* the distance (**A**) the ship continued to travel forward while he was in the air, and thus not in contact with it. (Figure 2-6b) However, when he jumps in the direction opposite the ship's travel, he appears to cover more than 2 meters. The 2 meters he would cover *plus* the distance (**B**) the ship continued to travel forward while he was not in contact with the deck. (Figure 2-6c)

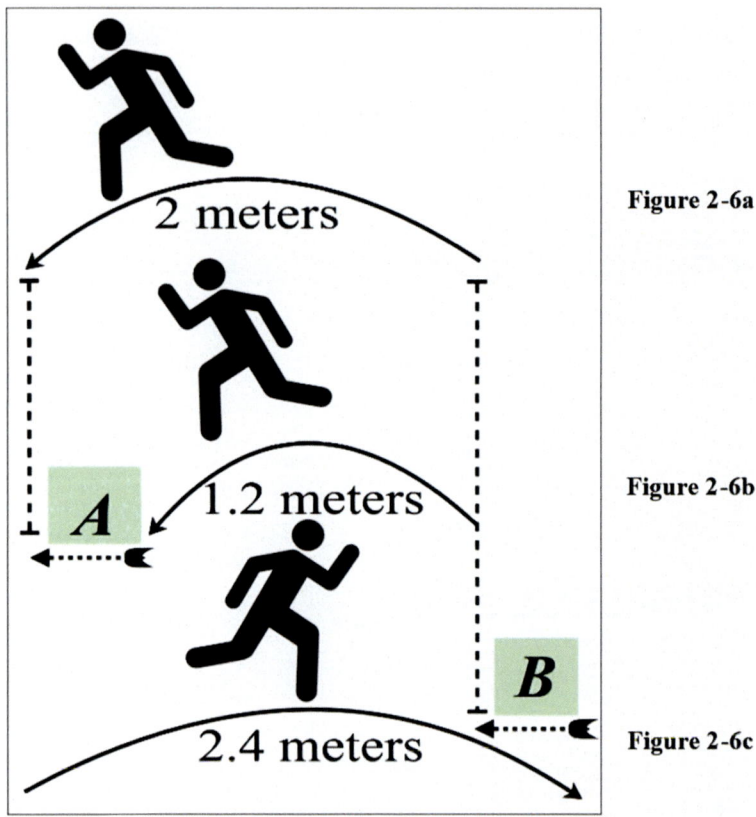

Figure 2-6a

Figure 2-6b

Figure 2-6c

Based on Aristotle's laws of motion, and <u>as measured on the ship's deck</u>, **Figure 2-6** shows the result of jumping within the hold of a ship. **Figure 2-6a** are the results when the ship is motionless. In **Figure 2-6b**, the ship travels in the same direction the sailor jumps. Length **A** represents the distance the ship continued to travel while he was in the air – thus not being pushed by the ship. Conversely, in **Figure 2-6c**, the length **B** represents the distance the ship continued to travel forward while the sailor was in the air, jumping toward the ship's stern.

Using the same logic, Galileo next considered the butterflies. Imagine several butterflies released inside the hold of a uniformly moving ship. As they flew about the hold, thus not in contact with the ship, they should immediately be smashed against the aft bulkhead. Finally, according to Aristotle, the dripping-water experiment onboard a uniformly moving ship should result in the drips missing the cup. Because as each drip falls through the air, it is no longer being pushed by the ship. Therefore, each drip should fall onto the deck *behind* the cup. Yet, are these results what we experience in everyday life? No. Notice Galileo's description and logical conclusion of these analogies.

SALVATI: *Shut yourself up with some friend in the main cabin below decks on some large ship, and have with you there some flies, butterflies, and other small flying animals.*

Have a large bowl of water with some fish in it; hang up a bottle that empties drop by drop into a wide vessel beneath it.

With the ship standing still, observe carefully how the little animals fly with equal speed to all sides of the cabin.

The fish swim indifferently in all directions; the drops fall into the vessel beneath; and, in throwing something to your friend, you need to throw it no more strongly in one direction than another, the distances being equal; jumping with your feet together, you pass equal spaces in every direction.

When you have observed all of these things carefully (though there is no doubt that when the ship is standing still everything must happen this way), have the ship proceed with any speed you like, so long as the motion is uniform and not fluctuating this way and that.

You will discover not the least change in all the effects named, nor could you tell from any of them whether the ship was moving or standing still.

In jumping, you will pass on the floor the same spaces as before, nor will you make larger jumps toward the stern than towards the prow even though the ship is moving quite rapidly, despite the fact that during the time that you are in the air the floor under you will be going in a direction opposite to your jump.

In throwing something to your companion, you will need no more force to get it to him whether he is in the direction of the bow or the stern, with yourself situated opposite.

The droplets will fall as before into the vessel beneath without dropping towards the stern, although while the drops are in the air the ship runs many spans.

The fish in the water will swim towards the front of their bowl with no more effort than toward the back, and will go with equal ease to bait placed anywhere around the edges of the bowl.

Finally the butterflies and flies will continue their flights indifferently toward every side, nor will it ever happen that they are concentrated toward the stern, as if tired out from keeping up with the course of the ship, from which they will have been separated during long intervals by keeping themselves in the air...

Galileo Galilei –
Dialogue Concerning the Two Chief World Systems

Because we don't know if Galileo ever conducted these experiments, these *thought experiments* are believed to have been the first to contradict Aristotle's laws of motion. But what makes Galileo's conclusions significant is that he correctly anticipated Newton's principle of *inertia*. And when the latter is coupled with the notion of relativity, it implies that any objects that share the same uniform velocity also share a uniform perspective. Or, in modern terms, they share the same *inertial frame of reference*. As a result, they will always appear motionless <u>relative to one another</u>. Let's again see this with the example of Galileo's ship.

Recall the jar of water suspended from the ceiling. Imagine watching as a single droplet falls toward a cup directly below. As it leaves the hanging jar, falling toward the cup, from the reference frame of all persons on the smooth sailing ship, the droplet *only* possesses the downward motion caused by gravity. (Figure 2-7a) Again, since everyone aboard the ship has the same horizontal motion (the ship's forward velocity), the net effect is that their shared movement cancels. The same principle applies to the Earth and everyone on it.

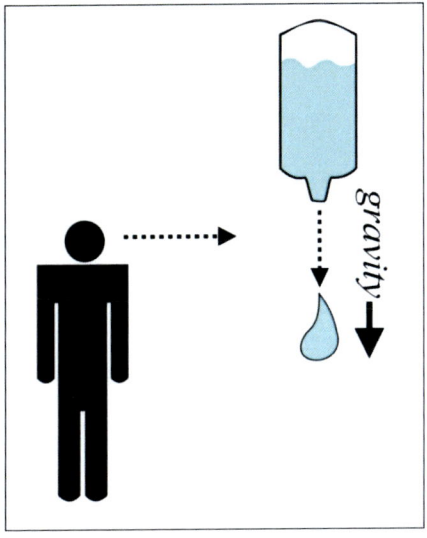

Figure 2-7a. Viewed from aboard the ship, within a shared inertial frame of reference, the drop of water will appear to fall straight down.

Even though the Earth is moving around the Sun, everything and everyone on the Earth shares that *motion*. Consequently, relative to one another, everything and everyone on the Earth that's standing still will appear motionless because they share the same inertial frame of reference. It's exactly like driving on the highway and looking over at a car traveling in the same direction beside you – that car appears motionless relative to you because you have matching speeds. But now, imagine someone who has a different perspective of Galileo's ship, someone watching the same drop of water falling, except this person is standing on the shore. How would the motion of the water droplet appear to them?

To someone viewing the droplet from shore, it would appear to possess the combination of <u>two</u> motions. (1) The downward motion due to gravity and (2) the forward motion it has gained from the moving ship. Therefore, *for the person located onshore,* who does not share the same inertial frame of reference as those on the ship, the dripping water will appear to follow a parabolic trajectory. (Figure 2-7b) Consider how revolutionary this concept was to those in the 17th century.

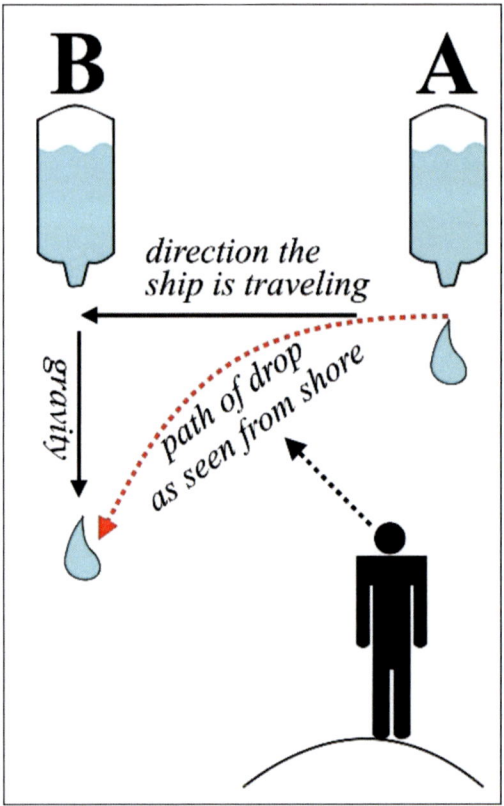

Figure 2-7b. When viewed from shore, from a relatively motionless inertial frame of reference, the drop of water's trajectory is perceived differently. It is not only falling due to the force of gravity, but it also shares the ship's forward motion. So, from the person's perspective onshore, the drop leaves the jar when the ship is at position **A**. However, when it reaches the deck, the ship (along with the droplet) has reached position **B**. Therefore, to an observer onshore, the drop of water appears to trace out a parabolic trajectory.

Two separate individuals viewing the same droplet would see its motion completely differently. The droplet is falling in a straight line to those <u>on the ship</u>. However, the same droplet has a curved trajectory to those <u>on the shore</u>. This thought experiment revealed that there was no such thing as absolute motion. Absolute motion excludes the idea that separate individuals can perceive an object's movement differently. Galileo's insight that motion was relative and depended on how it was viewed significantly advanced the theory of relativity. Yet, this was not the last of Aristotle's theories that Galileo would refute.

Some years after the publication of *Dialogue Concerning the Two Chief World Systems*, Galileo published his book *Discourses and Mathematical Demonstrations Relating to Two New Sciences*. This book disproved Aristotle's belief that heavier objects fell faster than lighter ones. Galileo realized that it was misleading to compare the speed at which a rock fell with a leaf. He understood that the air had a more significant effect on the less dense and greater surface area of the leaf, thus causing the latter to fall more slowly. Galileo, therefore, compared a falling heavy rock to an even heavier one.

We believe Galileo did conduct the following experiment by dropping the two rocks from the Tower of Pisa. Given that both rocks were "heavy," the effect of air resistance would be negligible on both. And when he dropped them, he found that they struck the ground simultaneously. Galileo had proven that, despite air resistance, all objects fall at the same rate. Then, he brilliantly reasoned how to logically disprove Aristotle's assumption that heavier objects fell faster than lighter ones.

SALVATI: If we take two bodies whose natural speeds are different, it is clear that on uniting the two, the more rapid one will be partly retarded by the slower, and the slower will be somewhat hastened by the swifter. Do you not agree with me in this opinion?

SIMPLICIO: You are unquestionably right.

SALVIATI: But if this is true, and if a large stone moves with a speed of, say, eight, while a smaller stone moves with a speed of four, then when they are united, the system will move with a speed of less than eight. Yet the two stones tied together make a stone larger than that which before moved with a speed of eight: hence the heavier body now moves with less speed than the lighter, an effect which is contrary to your supposition. Thus you see how, from the assumption that the heavier body moves faster than the lighter one, I can infer that the heavier body moves more slowly...

And so, Simplicio, we must conclude therefore that large and small bodies move with the same speed, provided only that they are of the same specific gravity.

Galileo Galilei –

Discourses and Mathematical Demonstrations Relating to Two New Sciences

Galileo's logic was irrefutable. And yet, there was an even more critical aspect to his experiment with falling objects. Again, consider Salvati's (Galileo's) logic:

***SALVATI**: Place a heavy body upon a yielding material, and leave it there without any pressure except that owing to its own weight; it is clear that if one lifts this body a cubit or two and allows it to fall upon the same material, it will, with this impulse, exert a new and greater pressure than that caused by its mere weight; and this effect is brought about by the [weight of the] falling body together with the velocity acquired during the fall, an effect which will be greater and greater according to the height of the fall, that is according as the velocity of the falling body becomes greater. From the quality and intensity of the blow we are thus enabled to accurately estimate the speed of a falling body. But tell me, gentlemen, is it not true that if a block be allowed to fall upon a stake from a height of four cubits and drives it into the earth, say, four finger-breadths, that coming from a height of two cubits it will drive the stake a much less distance, and from the height of one cubit a still less distance; and finally if the block be lifted only one finger-breadth how much more will it accomplish than if merely laid on top of the stake without percussion? Certainly very little. If it be lifted only the thickness of a* *leaf, the effect will be altogether imperceptible. And since the effect of the blow depends upon the velocity of this striking body, can any one doubt the motion is very slow and the speed more than small whenever the effect [of the blow] is imperceptible? See now the power of truth; the same experiment which at first glance seemed to show one thing, when more carefully examined, assures us of the contrary.*

Galileo Galilei –

**Discourses and Mathematical Demonstrations
Relating to Two New Sciences**

In other words, the longer an object fell under the force of gravity, the more its speed *increased* – thus the reason it struck with greater force. Aristotle had not considered the *full* implications of gravity being an *accelerating* force. But how could Galileo, lacking a radar gun, prove that all falling objects are accelerating? In a moment of inspiration, he came up with an ingenious solution using a ball.

Instead of dropping a heavy ball from the Tower of Pisa, Galileo realized that he could slow the ball's rate of descent by rolling it down an inclined plane. In both instances, the only force acting on the ball would be gravity. The only difference would be that the ball accelerating down an incline would do so at a *slower* rate of acceleration. (Figure 2-8) And it is much easier to track the acceleration of a ball rolling down an inclined plane than one falling from the Tower of Pisa. From conducting this experiment, what Galileo discovered next would change the course of physics forever.

Imagine once more the above ball rolling down an inclined plane. Here is the question: If you placed a second inclined plane opposite the first, how far up the second inclined plane would the ball travel? Now, this might seem like a trivial question, yet its answer has far-reaching consequences. And as we have been noting, the consequences reveal whether a theory is correct or not. Galileo found that if a second incline were placed directly after the first, the ball would climb the second to almost the exact height it had been released. In other words, the ball's motion was isochronous and harmonic.* Or, put another way, Galileo had discovered the law of conservation of momentum.

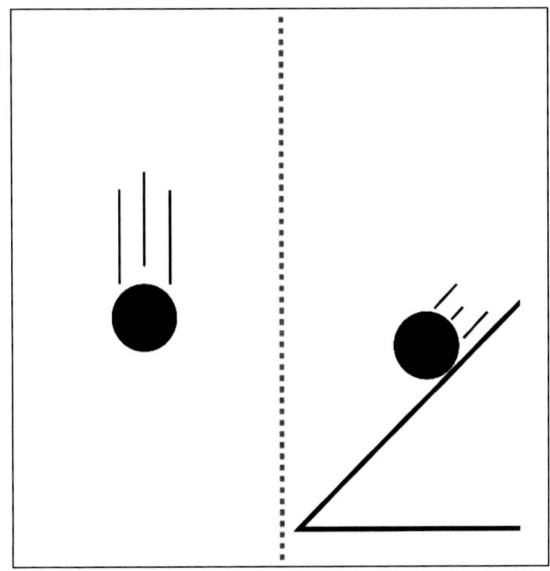

Figure 2-8. The ball on the left falls vertically very quickly. But the ball on the right, since it rolls down the inclined plane, is falling slower. Some of the latter's motion, and thus velocity, is being redirected horizontally. So, while both balls are equally affected by Earth's gravity, the ball on the right falls toward the ground more slowly.

---

* Galileo was one of the first to extensively study the motion of pendulums; he even envisioned the pendulum clock. He found that the time it took for a pendulum to complete its downswing was the same amount of time it took on its upswing. This is what is meant when it is said that a motion is "isochronous," meaning "of equal duration." Logically then, the concept of a harmonic oscillation must follow, meaning that the distance covered on the downswing must equal the distance covered on the upswing. Or, in simpler terms, whatever height the pendulum is released from, it will return to *almost* the exact height on its upswing. Galileo's next question was: Would these same principles work with a ball rolling down an inclined plane? The answer was yes. Under the force of gravity alone, a ball rolling down one inclined plane would take the same amount of time to roll up a second inclined plane. Furthermore, the ball would reach approximately the same height on the second inclined plane from which it was released on the first.

In Figure 2-9, the ball is released from a height of 1.5 meters. As it rolls down the first incline, it accelerates due to gravity. Once the ball reaches the bottom, it has *fallen* as far as it can and has, therefore, gained all the momentum that gravity can supply. However, when the ball encounters the second incline, it will start to climb it. But gravity would now be acting in the opposite direction of the ball's travel, causing the ball to *decelerate*. So, in asking how far up the second incline the ball will travel, we only need to answer the question: From what height did it fall? Since the ball was released from a height of 1.5 meters, gravity supplied *1.5 meters* of momentum. Thus, as it climbs the second incline, gravity will subtract *1.5 meters* of momentum. It will reach the same height from which it was released. Indeed, this would occur even if the second incline were at a different angle, and this last fact is what led to the profound truth.

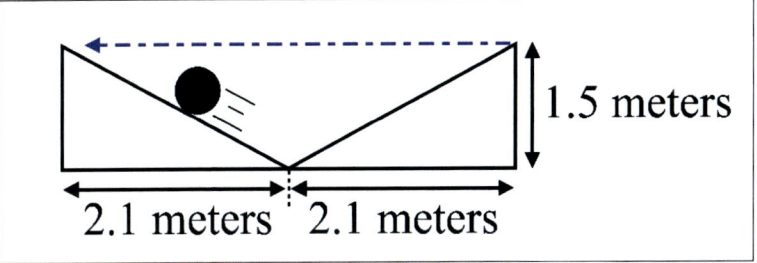

Figure 2-9. When a ball is rolled down an inclined plane from a specific height, it will rise to the same height on a second inclined plane. In other words, its motion is isochronous and harmonic.

Imagine that we begin to lower the angle of the second incline incrementally. As we did so, the ball would have to travel a bit farther horizontally to reach the vertical height of 1.5 meters. (Figure 2-10) Now, put yourself in Galileo's position and follow this experiment to its logical conclusion. The more the second incline's angle is reduced, the longer (horizontally) the second incline must be to reach a height of 1.5 meters. Thus, the farther the ball must travel horizontally. So, what happens when the angle reaches 0 degrees? Theoretically, once the angle is 0 degrees, the ball *should* roll along the horizontal plane indefinitely.[*] (Figure 2-10) Again, Galileo was, unknowingly, anticipating Newton's first law, the law of inertia – an object in motion will remain in motion unless it is acted upon by another force.

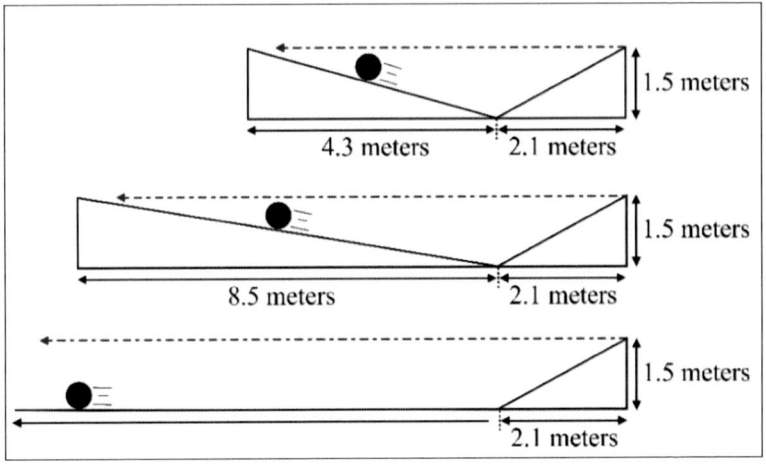

Figure 2-10. From whatever height the ball is released, it will return to that height on the other side, even if the angle of the incline is different. And Galileo correctly deduced that the ball should also (neglecting air resistance and friction) continue to roll horizontally for an indefinite period if there were no opposing incline.

All the above conclusions were significant steps forward in developing the theory of relativity. Each also refuted one of Aristotle's theories. And still, for all his scientific insights, Galileo was not as perceptive when it came to people. Or, more specifically, in his anticipating the church's reaction to his challenging of officially endorsed church doctrine. Galileo thought his reputation and powerful connections would somehow protect him from censure. However, he made the mistake of making his dispute with the church public. And his public questioning of church dogma is what eventually cost him his freedom. In 1633, Pope

---

[*] In practice, both friction and air resistance would eventually cause the ball to stop rolling. But in a vacuum and under frictionless conditions, the ball would continue to roll indefinitely along the horizontal plane.

Urban VIII summoned Galileo to Rome to recant his work. Upon his arrival, Galileo was arrested, tried, and convicted of violating an injunction on the teaching of Copernican views. Nevertheless, it was too late; the "damage" had already been done.

Galileo had conclusively proven that, contrary to Aristotelian physics, motion was not a process[*] that occurred to mass. Instead, motion is a state[†] in which mass exists. And although he didn't understand why, Galileo had also proven that all objects, no matter their weight, fell at the same rate. Both the above ideas were vital in preparing the way for a more thorough explanation of gravity.

Galileo's achievements changed the course of civilization, and it might have appeared that some time would pass before his deeds would be surpassed. But, most fittingly, in the same year as Galileo's death, the next contributor toward the theory of relativity was born. And this next scientist is recognized as the greatest physicist in history: Sir Isaac Newton. However, before we consider Newton's contributions to relativity, let's examine a crucial mathematical debate that was raging. And this debate would lead to the inclusion of the only aspect of special relativity that had yet to be considered – energy.

---

[*] A process is defined as a systematic series of actions directed toward some end. By Aristotle's definition, "motion" was a process. Because if an object *only moves when it is pushed*, then the instant the force begins to *push* the object would be the start of the process. And so long as the force acts on it, the process continues, and thus, the object continues to move. However, once the force is removed, the process immediately stops. And according to Aristotle, the object would then return to its position of absolute rest.

[†] A "state" is defined as the particular condition that someone or something is in at a specific time. Galileo showed that motion was relative. An object that may appear to be standing completely still to one person may, to a second individual, be moving at uniform velocity. Thus, an object's state can either be one of uniform motion or rest; Galileo's anticipated Newton's 1st law of motion.

Chapter 3

René Descartes was born in 1596. He held a bachelor's degree in Law from the University of Poitiers. Yet, despite studying law, his passion was physics, philosophy, and mathematics, which included the works of Galileo. Today Descartes is primarily remembered for his creation of analytical geometry and for the philosophical axiom: *Cogito, ergo sum* ("I think; therefore, I am"). However, there is another, lesser-known fact regarding Descartes. For the whole of his professional life, he and other celebrated

René Descartes (1596–1650)

contributors to science were engaged in a very heated, metaphysical debate. The debate centered on one question: What exactly is a force? [*]

Another idea promoted by Aristotle was that of *potentiality* versus *actuality*. These terms were meant to be two types of opposing forces. Those forces that had "potential" (or that was possible) he called *potentiality*. Those forces in the process of occurring, he called *actuality*. Over time, both terms were eventually renamed, and each came to be associated with particular mathematical formulas. The first term became known as the "living force" (*vis viva*), and its formula was $F = mv^2$. The second term became known as the "dead force" (*vis mortua*); it was described by the formula $F = mv$. However, by the 17th century, it was realized that only one of these mathematical terms was correct. As a result, since only one of these could rightly be called a "force," it didn't take long before scientists and mathematicians began to take sides.

Each term came to have prominent supporters and detractors. Descartes advocated the "dead force." Still, he and other "dead

---

[*] Along with Descartes, this dispute involved other famous scientists such as Christiaan Huygens and Gottfried Wilhelm Leibniz. To a lesser extent, it also involved Samuel Clarke, Leonhard Euler, Jean and Daniel Bernoulli, and Émilie du Châtelet. Even Isaac Newton was drawn into the debate.

Gottfried Leibniz (1646–1716)

force" advocates were obliged to admit that there existed situations where their formula didn't apply. The lead advocate of the "living force" was Gottfried Leibniz, the man who independently, though secondly, conceived of calculus. Yet, like the formula used for describing the "dead force," that used to describe the "living force" had its limitations. Making the situation even worse, when Isaac Newton proposed his laws of motion, he further complicated the issue when he introduced a third equation, $F = ma$. Newton defined a "force" as a mass multiplied by its *acceleration*. The debate eventually became so intense that it was soon divided along national boundaries – the English siding with Newton, the French with Descartes, and the Germans with Leibniz. However, the actual dilemma was they were all using the same <u>word</u> (force) to describe different things.[*]

The formula $F = mv$ is today known as momentum. This term is defined as the measure of an object's motion. As already mentioned, an essential concept concerning momentum is that it is *always* conserved. So if one object is losing momentum, there <u>must be</u> another gaining it. A second important characteristic of momentum is that it is a vector quantity. Vectors have magnitude (i.e., amount) *and* a direction of application (e.g., left, right, up, or down). Therefore, momentum is not a force, which is why we represent it today with the Greek letter rho; momentum's formula is now written as $\rho = mv$.

In contrast, the formula $F = mv^2$ measures the *capacity* to move an object; this second, <u>though incomplete</u>, formula is what we now call *energy*. Energy is a scalar quantity, meaning it *only*

---

[*] The actual debate centered on how the universe was created. For example, was the universe created using a "living force" (i.e., once the energy required to create the universe was supplied, the universe became self-sustaining, no longer requiring any more energy input)? Or did God use a "dead force" to create the universe (i.e., once the energy required to create the universe was supplied, the universe began to degrade, meaning that, in time, God would have to supply more energy for the universe to continue)?

measures magnitude; the direction of application is not of concern. For example, the energy required to push a 50-kilogram rock on a flat surface is the same whether you push the rock to the left or right. The same amount of energy is required regardless of its direction of application; more on this formula later. Consequently, in the end, the person who won the debate was Isaac Newton.

Newton's formula, $F = ma$, is the accepted definition of a *force*. It describes how the velocity of an object *changes* when forces are applied to it. Newton's formula is also a vector quantity.[*] But again, the primary problem was that each side was trying to use the same word (force) to describe entirely different phenomena. The person who eventually ended the debate was the mathematician and philosopher Jean le Rond d'Alembert. He did so in his 1758 revised treatise *Traité de Dynamique*. But before he did so, let's consider the important work of the sole woman in this list of historical contributors to the theory of relativity.

Émilie du Châtelet was born in Paris in 1706 and wasn't the average 18th-century woman. While a child, she showed a natural aptitude for science and math – subjects considered inappropriate for a proper French lady. As a teenager, she excelled at games of chance. She would use the money she won to purchase books and lab equipment to conduct experiments. And as she neared marrying age, her father is recorded as lamenting, "My youngest flaunts her mind and frightens away the suitors." All of this gravely vexed her mother. Yet, in an ironic twist, while her mother repeatedly threatened to send her away to a convent, it was her father who nourished her talents. He even arranged for her to discuss

Émilie du Châtelet (1706–1749)

---

[*] Momentum differs from a "force" in that momentum is based on uniform (unchanging) velocity. Force is based on acceleration, the latter being defined as the rate of change (speeding up or slowing down) per unit of time. The following is a simplified way to remember the difference. Momentum tells you how much motion an object has; force tells you how hard you'll have to work to change the object's momentum.

scientific theories with prominent astronomers. It was fortunate for du Châtelet that her father held a high-ranking position in King Louis XIV's court, which allowed her to achieve what other women would not experience as a right for many decades.[*] But, returning to the living/dead debate: should the velocity (v) be multiplied by the mass (mv), or should it be *squared* and then multiplied by the mass (mv^2)? Émilie du Châtelet helped to provide an important part of the answer.

To help settle the issue, du Châtelet designed an apparatus and conducted an experiment that consisted of dropping a lead ball from increased heights onto incredibly soft clay.[†] (Figure 3-1[‡]) As the height that the ball fell from increased, there was a corresponding increase in the ball's velocity. Equally important, this increase in velocity was accompanied by an increase in penetration depth. And, since the ball's mass remained constant, the depth that it penetrated the clay could only be a function of its velocity. So, what were the results of du Châtelet's experiment?

After establishing a baseline, du Châtelet found that the penetration depth was four times as deep when the ball's velocity was doubled ($2^2 = 4$). When the velocity was tripled, the penetration was nine times as deep ($3^2 = 9$), and so on. Thus, the

---

[*] Émilie du Châtelet was a child prodigy in the real sense of the word. She not only excelled at science and math but spoke six languages by the age of 12, played the harpsichord, sang opera, and authored multiple scientific books and papers. But, to appreciate who she was, you must examine the final several months of her life. At the age of 42, du Châtelet became pregnant, which often resulted in death for a woman in the 18th century. Being aware of this, she worked tirelessly, sometimes as many as 18 hours per day, to complete a French translation with commentary on Isaac Newton's Principia Mathematica. She even included her notes, examples, derivations, and experiments to confirm Newton's theories. Her determined effort paid off; she completed her book just days before she gave birth and died. To this day, her translation of Principia Mathematica is still the standard translation into French, a testament to her effort.

[†] Willem 'sGravesande was a Dutch mathematician and philosopher. He was the first to conduct the experiment, which he later communicated to du Châtelet. As is still done today, his results needed to be "peer-reviewed" before being accepted. Du Châtelet successfully repeated and confirmed the experiment. Indeed, her results became the definitive and cited verification.

[‡] Image credit: University of Hamburg

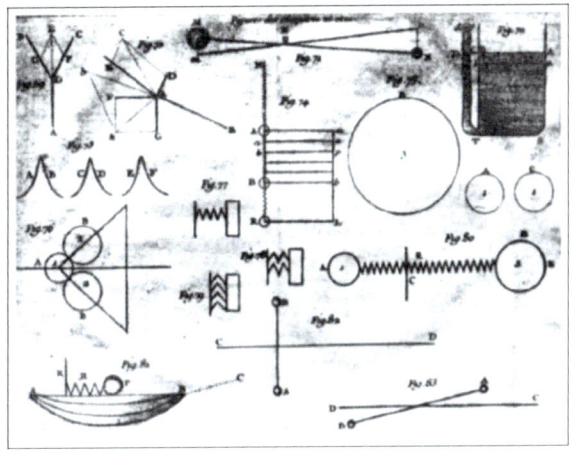

Figure 3-1. The actual technical drawings of the apparatus that du Châtelet constructed to independently verify Leibniz's conclusions.

force (or energy) that a falling object possessed was determined by its mass multiplied by the velocity *squared*. This confirmation turned out to be of immense importance to d'Alembert, who indisputably resolved the debate a few years later.

Jean le Rond d'Alembert (1717–1783)

D'Alembert's insight came when he re-evaluated the formula $F = mv^2$. He realized that this formula, initially devised by Galileo, was not in its final form; it needed additional refinement to the more accurate $F = \frac{1}{2} mv^2$. This small adjustment would prove invaluable to Michael Faraday and James Maxwell several decades later. While conducting unrelated research, the latter men independently found that d'Alembert's formula could also be derived in the field of electrodynamics, a field that dealt exclusively with electromagnetic *energy*. And this final insight expanded the application of d'Alembert's formula. It was now seen as a description of all types of energy.[*] For example, when applied to

---

[*] In physics, energy is defined as the capacity to move a mass through a distance. In simpler terms, it is the capacity to do work.

motion, as in du Châtelet's experiments, d'Alembert's formula calculates kinetic *energy*.[*] But why was this debate over formulas even necessary?

For those of us who aren't so mathematically inclined, this debate over a living/dead force may seem to be mere semantics, an argument over words. However, it was a crucial part of the development of the theory of relativity. Recall that one aspect of special relativity is Einstein's formula $E = m\mathbf{c}^2$. The quantity $\mathbf{c}$ in the formula represents the speed of light. Like the velocity (v) in the formula for kinetic energy, both represent velocity, and both must be *squared* to determine their final answers.

Still, d'Alembert's reformulation of the equation for kinetic energy did not occur until the middle of the 18th century. Returning to our narrative, the start of the 17th century was a significant turning point in English history. The century began with the death of Elizabeth I on March 24, 1603. Then came the "Union of the Crowns," when James VI, King of the Scots, became king of England. England and Scotland were now united under a single monarchy, and James VI of Scotland became James I of England. Yet, James' accession, and the uniting of these lands, also came with years of instability.

James I died in 1625 and was succeeded by his second son, Charles I of England. Charles shared his father's dream of fully uniting the territories – England, Scotland, and Ireland – under his authority. However, because Charles I insisted that his authority was divinely ordained, parliament was wary of granting him more power. Eventually, after more than 17 years of tension, things reached a critical point when the English Civil War erupted in 1642. Intermittent clashes between the supporters of Charles I and those of the parliament continued for the next nine years.

Finally, on September 3, 1651, parliamentary forces defeated those of Charles II at the Battle of Worcester. This battle changed the course of the world. For the first time in European history, a monarchy was secondary to a legislative body. However, England endured much more than a civil war during this period; the

---

[*] Kinetic energy (KE) is defined as the energy of a body (or a system) with respect to its *motion*. Its formula is written as $KE = \frac{1}{2} mv^2$.

bubonic plague also afflicted the country. The largest outbreak is known as the "Great Plague of 1665." But it was during this period of civil war and sporadic plague that the next contributor toward the theory of relativity was born.

Sir Isaac Newton (1643–1727)*

Isaac Newton's father, also named Issac, was born in 1606. It's known that he owned a fair amount of land and animals, which probably meant that he was not impoverished. Yet, he was illiterate. At the age of 36, he married Hannah Ayscough. Aside from the above, not much else is known about Newton Sr. Sadly, he never knew his son or witnessed his many scientific feats. Just six months after getting married, Newton Sr. died.

On the other hand, Hannah was born in 1623 in the village of Market Overton, and she came from a very well-educated family. Hannah's brother, William, earned a master's degree from Cambridge in 1637. Hannah Newton prematurely gave birth to Isaac Newton Jr. just two months after her husband's death. It's said that, at birth, Newton was small enough to fit into a quart-sized measuring tin. In fact, the newborn Isaac was so weak that nurses sent to retrieve medicine, sure he would not be alive upon their return, took their time to rest. But baby Newton did survive.

When Newton was three years old, his mother remarried, he was a minister by the name of North Witham. Leaving her young child in the care of her parents, James and Margery Ayscough, Hannah moved to live with her new husband, approximately 1.6

---

* Some sources list Newton's birthday as December 25, 1642, while others list January 4, 1643. The difference lies in which calendar is consulted. Continental Europe had already adopted the Gregorian calendar; however, England used the Julian calendar until 1750. Besides having the New Year beginning on March 25 instead of January 1, the Julian calendar also ran ten days behind the Gregorian calendar. Therefore, according to the Julian calendar, Isaac Newton was born on December 25, 1642. In contrast, according to the Gregorian calendar, which we use today, he was born on January 4, 1643.

km away. And though she wasn't that far away (in terms of distance), it appears she had little to no contact with her first child, which took a tremendous emotional toll on the young Newton. By all accounts, Newton did not have a happy childhood. As a result, he developed a very deep-rooted hatred for both his mother and stepfather, which, in turn, caused him many feelings of guilt. So it's not surprising that these early events of Newton's life left his scholastic development in serious jeopardy.

By 1653, both Newton's grandfather and stepfather had died. Thus Hannah, now in possession of some wealth, returned to her parent's home with Newton's half-brother and two half-sisters. Newton's school reports from the period described him as being idle and inattentive. So, with her eldest son's academic future in doubt, Hannah decided to remove him from school to manage her estate. But he showed even less aptitude for this. Therefore, by the time he was 17, Newton's entire future had become uncertain. Still, while most could not see his potential, his uncle William Ayscough did. Consequently, at his uncle's suggestion, in 1661, Newton returned to school, he enrolled in Trinity College, Cambridge.

Despite enrolling as a law student, the curriculum still required that he study the works of Aristotle, Descartes, and Boyle. Newton was also introduced to Copernicus' astronomy, Kepler's mathematics, Euclid's geometry, and Galileo's mechanics. These topics whetted Newton's appetite for physics. Thus, by 1663, he had changed his major, going from law to mathematics. In April of 1665, Newton received his bachelor's degree in science, though without honors or distinction. So even by this late date, Newton had yet to display the genius for which history would remember him – that is until the Great Plague.

It's estimated that, during the two years of the Great Plague, some 68,000-100,000 persons died in London. Throughout England, the death toll probably topped 200,000. To slow the disease's spread, many of the city's population retreated to the countryside. Newton, too, departed Cambridge and spent two years in Lincolnshire. During this period of social distancing, Newton developed most of his theories, including calculus. And the extent to which these discoveries changed the world cannot be overstated. So, knowing a bit of his life story, let's now consider how Isaac Newton added to the theory of relativity. Moreover, if this story is going to be told properly, it can't be done without mentioning the most famous *falling* apple in all history.

Newton's Apple

No explanation of Newtonian mechanics would be complete without recounting the story with which Isaac Newton has become synonymous. It's said that he described the incident many times. However, the story's details were not written down until a year after his death. One of Newton's close friends, William Stukeley, described the event as follows:

> *After dinner, the weather being warm, we went into the garden and drank tea, under the shade of some apple trees. [Newton] told me, he was just in the same situation, as when formerly, the notion of gravitation came into his mind. It was occasion'd by the fall of an apple, as he sat in contemplative mood.*
>
> *'Why should that apple always descend perpendicularly to the ground,' thought he to himself. 'Why should it not go sideways, or upwards? But constantly to the earth's centre? Assuredly, the reason is, that the earth draws it. There must be a drawing power in matter.'*

William Stukeley –

Memoirs of Sir Isaac Newton's Life

Assuming the story is true, as Newton watched an apple fall from a tree, it gave him a singular moment of clarity. It allowed him to discover the source of gravity – matter itself. Just as certain metals can generate a magnetic field, Newton reasoned that all matter generated a gravitational field. This latter field was the cause of attraction between material objects. And the larger the mass of an object, the larger its gravitational field. Forming this connection between mass and gravity, Newton's laws of motion must necessarily follow, beginning with the first:

Newton's 1st Law of Motion:

An object in uniform motion (or at rest) will remain in uniform motion (or at rest) unless another force acts upon it.

This first law is also called the law of inertia. Let's see if we, watching the same apple fall from the tree, can reach the same conclusion.

While the apple is in the tree, all forces are perfectly balanced. Put another way, the force of gravity pulling the apple down is canceled by the branch holding the apple up. Therefore, because the forces acting on the apple are equal and opposite, it behaves as if no forces are acting on it. It's as if the apple were sitting motionless in space. And as long as these forces remain balanced, the apple will sit in the tree indefinitely. Thus, an object at rest will remain at rest. But once the apple ripens, the tree begins to loosen its grip. Finally, at some point, the force of gravity becomes greater than the branch holding the apple. It is at that moment that the apple falls. However, the real question becomes: What is happening *as* the apple is falling?

As the apple falls, it doesn't stop in midair, then turn left and hover for a moment before continuing to fall toward the ground. No, it falls directly to the Earth's surface, which leads to another question. For the brief moment that the apple falls toward the ground, what would happen if the Earth, and everything around it, instantly disappeared? Well, it is the Earth that's generating the gravity that's pulling on the apple. So, without the Earth (or any other force) to act on it, the apple would continue moving through space at whatever speed it was moving before the Earth disappeared. In other words, an object in motion will remain in motion unless acted upon by another force.

Therefore, Newton's conclusion agreed with Galileo, which was the exact opposite of Aristotle's. Objects *are not* at rest until they are pushed. Instead, whatever state an object is in, even if in motion, it will remain in that state until some force acts on it. Thus, we've reached the same conclusion as Newton in his first law of motion. Indeed, we were able to do so by merely considering a falling apple. Now, what about the second law?

Newton's 2nd Law of Motion:

$$F = ma$$

Galileo had shown that all objects near the Earth's surface fall at the same rate. Therefore, ignoring air resistance, an apple will fall to the ground at the same velocity as a cannonball dropped from the same height. Yet, as Galileo had also shown, when these objects reach the ground, they strike with vastly different amounts of force. What do both objects have in common that causes them to fall at the same velocity? Yet, what differences do they possess that cause them to strike the ground with different amounts of force? The answers are the Earth and the object's masses, respectively.

Both the apple and the cannonball are accelerated equally due to the Earth's gravity; gravity accelerates everything at the same rate at the Earth's surface. Therefore, the only difference is their respective masses. An apple is about 135 grams. In comparison, a typical cannonball has a mass of about 200 000 grams. As a result, even though they will fall at the same rate, the cannonball will strike the ground with a greater *force* than the apple. So, what did Newton conclude? Since there are only two variables involved (the object's mass and the Earth's gravity), these two variables combined must be what defines a force. In other words, the force required to change an object's state – from falling through the air to resting on the ground – equals the object's mass (m) multiplied by how fast it was falling (accelerating): $F = ma$. Again, imagine the apple while still in the tree.

For the apple's state to change from motionless (while in the tree) to falling through the air, it must experience acceleration; no object can go from 0 to 60 without undergoing acceleration. Similarly, an object in motion can't speed up, slow down, or turn left or right without some type of force being applied.* Imagine, for instance, an object moving in space in a straight line. If a force is applied to the object *in its direction of travel*, it will speed up. And when you remove the force, the object will have attained a new *uniform* velocity. A force applied *in the opposite direction of*

---

* Changing direction is considered acceleration.

travel will cause the object to experience a negative acceleration (i.e., decelerate). But, again, once the force is removed, it will settle into its new state of uniform velocity (or a state of rest). Therefore, in the previous example of the apple and cannonball, each has its respective mass (*m*), and gravity supplies the acceleration (a). Consequently, the *force* they strike the ground with (changing their states from one of motion to a state of rest) is calculated using the equation F = *m*a.

As with his first law of motion, Newton's Second Law also contradicted Aristotle. Remember, Aristotle held that a falling object's speed was proportional to its mass, the force applied to it, and inversely proportional to the medium it was falling through. Essentially, how fast an object fell was determined by how *heavy* it was, how hard it was *pushed*, and how *dense* the air was. But Newton's Second Law of motion stated that the density of the medium was immaterial. The object's mass and acceleration played the primary roles, and these two are the only variables required to define a force.*

Newton's 3ʳᵈ Law of Motion:

When one body exerts a force on a second body, the second body simultaneously exerts a force (equal in magnitude and opposite in direction) to that of the first body.

Or, more informally, this law is usually stated as: For every action, there is an equal and opposite reaction.

Recall Galileo's experiment with the inclined planes; it demonstrated that momentum was always conserved. Similarly, while Newton's apple was motionless in the tree, its momentum was conserved; the force pulling the apple toward the Earth was balanced by the force of the branch holding it up. However, the

---

* However, this is not to say that the presence of a medium, such as air, doesn't affect falling objects. The key point is that the medium only indirectly affects the rate that an object falls. In the absence of a medium, all objects fall at the same rate. This conclusion was experimentally verified in 1971 by the Apollo 15 moon crew.
https://moon.nasa.gov/resources/331/the-apollo-15-hammer-feather-drop/

instant the apple began to fall, what equal but opposite force was acting to satisfy Galileo's conclusion of conservation of momentum? It isn't easy to envision when we view the situation in this way. So let's examine it differently.

Imagine the reverse, you picking up the apple from off the ground. You impart an accelerative force opposite to the Earth's as you do so. And this satisfies the conservation of momentum. There are two opposing forces (you and the Earth), with the apple caught in the middle. But what equal yet opposite force occurs as the apple falls toward the ground? As he pondered this circumstance, Newton made another fantastic mental leap.

If gravity radiates from all matter, then as the Earth pulls the apple downward, the apple must – since it is also composed of matter – be pulling the Earth upward. Thus, we arrive at Newton's Third Law of motion: For every action, there is an equal and opposite reaction. This statement is the key to conserving momentum, and it defines Newton's Law of universal gravitation.[*] The Earth pulls on the apple, causing the apple to fall toward the ground. However, the apple is pulling on the Earth at the same time. The only difference is that the Earth's mass is exponentially many times greater. Still, although the force of the apple pulling on the Earth is slight, it does, nonetheless, exist! Therefore, momentum is conserved. These are Newton's three laws of motion. But why were these laws so exceptional? Well, first, they explained how the Moon orbited the Earth.

Consequences of Newtonian Mechanics

As you may have noticed, necessary consequences follow each time a new theory is proposed. We discern if a new theory is accurate by exploring its consequences. And Newton's laws of motion definitely implied enormous consequences. Let's see what these were by imagining the apple falling from the tree again.

---

[*] Newton's law of universal gravitation states that every object in the universe that possesses mass attracts every other object having mass.

So, the apple is about to break free of its branch and fall. Now, imagine a person standing beside the tree holding a second apple. The instant the first apple breaks free of the branch, the person standing beside the tree throws the second apple from the same height and perfectly horizontal. Which apple will hit the ground first? Well, besides the second apple having a horizontal velocity, the force gravity exerts on both remains equal. Therefore, both apples would fall toward the ground at the same rate. (Figure 3-2) Of course, as in Galileo's ship experiment, the second apple's

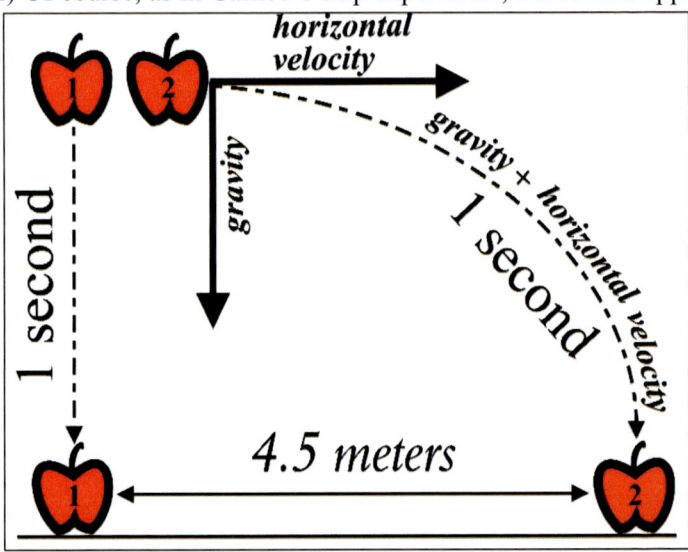

Figure 3-2. Even though the second apple has a horizontal velocity, gravity still causes it to fall toward the Earth at the same rate as the first apple. The result on the second apple is that, while hitting the ground at the same instant as the first, it will do so 4.5 meters away.

horizontal motion, combined with gravity's vertical pull, would cause the second apple to travel through the air with an arcing trajectory. Still, both apples would reach the ground simultaneously. The only difference is that the second apple would land a few meters away. But what would happen if a professional baseball pitcher were to throw the second apple?

A professional pitcher can throw an apple at around 145 km per hour. Still, so long as he throws it perfectly horizontal, the outcome would be similar to the example above. The only difference would be that the second apple's greater momentum would carry it even farther; it would trace a larger arcing path toward the ground. What would happen, though, if the apple were fired from a cannon? Well, assuming the second apple remained intact, it would have an even greater horizontal velocity. Yet, just as in the previous examples, its horizontal velocity would not affect the force of gravity acting on it. Therefore, its increased horizontal velocity, combined with gravity, would simply trace an even larger arc as the second apple fell toward the ground. (Figure 3-3) Can you now begin to see the consequence this has on the Moon orbiting the Earth? Newton did.

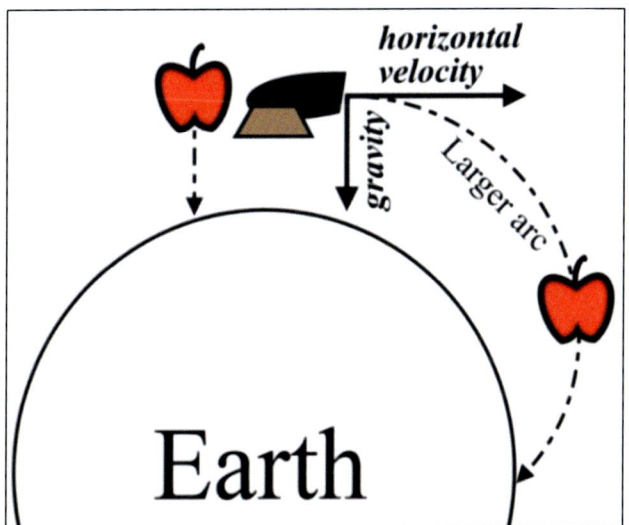

Figure 3-3. The second apple is given an even greater horizontal momentum. Still, gravity causes it to fall toward the ground at the same rate as the first apple, just along a more significant, arcing path.

The more horizontal momentum you impart to the apple, the larger the arc it traces out before hitting the ground. Remember, though, that the Earth is round. Thus, there exists a horizontal velocity that the apple can be thrown such that the arc it traces out will <u>match</u> the Earth's curvature precisely. And what happens when this velocity is achieved? The apple, because its trajectory now matches the curve of the Earth, *will never hit the ground*. In other words, the apple will have achieved a stable orbit. This same principle applies to the Moon.

The Moon has a large horizontal momentum (i.e., it moves at a high horizontal velocity). And when this velocity is combined with the downward pull of the Earth's gravity, it causes the Moon to trace out a curved trajectory equal to the curvature of Earth. In other words, the Moon is falling toward the Earth. But because of its high horizontal velocity, it *misses* hitting the Earth on each orbit. (Figure 3-4) This logic is the same used to place satellites in

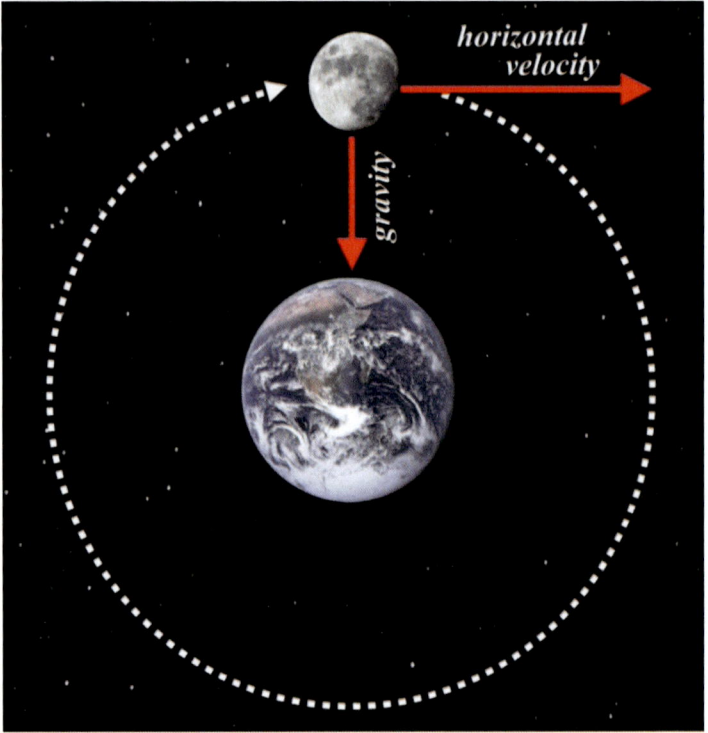

Figure 3-4. The Moon is falling toward the Earth; however, its large horizontal momentum (orbital speed of 3680 km/hr) allows it to follow a trajectory equal to the Earth's curvature. Therefore, the Moon misses hitting the Earth on each orbit.

orbit. Give any object within the Earth's gravitational field enough horizontal momentum, and there exists some height at which the path it traces will equal the Earth's curvature. We will then say that (at that height and velocity) the object has achieved a stable earth orbit.* Yes, Newton's laws of motion explained how the Moon orbited the Earth, a fantastic achievement in human thought. However, although exceptional, Newtonian mechanics had other, more controversial after-effects.

Newton's concept of universal gravitation was as a force that radiated in all directions from matter. Just as significant, he thought it extended indefinitely into space, acted *instantaneously* in all directions, and thus affected every part of the universe at once. These ideas about gravity led to some very significant consequences. For example, it suggested what we today call *action-at-a-distance*.

Today, we know that the Earth is 150 million km from the Sun and that light takes about $8^1/_2$ minutes to reach us. Therefore, if the Sun were to disappear right *now*, it would take roughly $8^1/_2$ minutes for us to know it. However, this was not what Newton thought would happen. Recall that, like the Moon, the Earth's movement around the Sun is the combination of two motions: The Sun's gravity pulling the Earth toward it and the Earth's horizontal momentum. So if Newton's idea of universal gravitation were true, what would happen to the Earth if the Sun were to disappear? Since the latter's gravitational pull would *immediately* disappear, the only motion left would be the Earth's horizontal momentum. Thus, the Earth would *instantly* race off in a straight line into space. But this conception of instantaneous gravity created several logical problems.

All human experience teaches us that events occur only *after* contact is made. It takes a breeze to ruffle a flag, a thrown ball to break a window, and a pebble to ripple a pond. Yet, the idea of physical contact means that time is involved. A flag does not wave before the breeze reaches it. A window does not break before the

---

* Of course, a stable orbit cannot be achieved within the Earth's atmosphere. The presence of air causes friction that slows moving objects, thus causing them to fall to the ground. This same friction is experienced by orbiting objects attempting to return to Earth. As an object enters the Earth's atmosphere at a high velocity and is slowed due to friction (i.e., air molecules colliding with the object), it gets extremely hot. Indeed, most objects entering the Earth's atmosphere literally burn up due to friction. But objects outside the Earth's atmosphere (like the Moon) experience no friction due to the lack of air. They can, therefore, orbit the Earth indefinitely.

ball strikes. And we have never seen the pond begin to ripple before the thrower releases the pebble. But this was how Newton thought gravity behaved. Newton's concept of gravity was of a force that was both universal and instantaneous, meaning that the gravity that radiated from every particle of mass in existence was *universally* pervasive. Therefore, the instant there was any change in mass – in any gravitational field (anywhere) – the entire universe was *instantly* affected. To justify his belief, Newton borrowed an idea from Aristotle: the notion of a universal aether.

Newton believed that a universal substance called the aether existed, and this was the substance that transmitted gravity. It was like a tube lined with marbles: The moment you insert a marble into one end, a marble immediately exits the opposite end. Newton thought that this aether permeated the entire universe, making instantaneous transmission of gravity possible. However, the inclusion of the aether as the transmitter of gravity only added to the flaws in Newtonian mechanics.

In promoting the idea of the aether, Newton was attempting to support the concept that space and time were *absolute*. Indeed, he explicitly mentioned his belief in this. Thus, Newton's explanation of gravity and his belief in action-at-a-distance makes perfect sense when viewed from this perspective. Yet, in advocating these ideas, he ignored the consequences that resulted. We can illustrate the problem by recalling the time it takes for light to reach us from the Sun.

It takes light $8^{1}/_{2}$ minutes to reach the Earth from the Sun and 12 minutes to reach Mars. However, because Newton didn't accept the notion of an observer's frame of reference, if the Sun were to disappear right now, it would do so for everyone, everywhere in the universe at the same time. So, anyone looking at the Sun when it disappeared would see it do so simultaneously, whether you were located on the Earth, Mars, or in the Andromeda galaxy. And this nicely illustrates how Newton thought gravity behaved. Going one step further, he didn't even regard space and time as *real*, in the sense that a flower or a rock exists. Therefore, Newton couldn't imagine a physical world directly *interacting* with space and time. Hence, belief in absolute space and time formed the cornerstone of Newtonian mechanics. And the idea of action-at-a-distance necessarily followed, though not everyone who examined his theories accepted this conclusion. Gottfried Leibniz (the independent creator of calculus) stated that space and time would make no sense were it not for the existence of <u>motion and objects</u>. To understand what Leibniz meant, consider the following example.

Let's assume that you existed all alone in the universe. Does space exist? That would be difficult to answer since we can only conceive of "*space*" when at least two objects exist separate from each other. For instance, imagine a wall to your left that is 1.5 meters away. Therefore, between you and the wall is 1.5 meters of space. But if you were all alone in the universe, with nothing else to gauge distance by, the concept of *space* becomes meaningless. Now, let's say two objects exist in the universe: you and a single rock located 3 meters apart. While you can now conceive of space, if neither you nor the rock has any relative, repeating motion, how do you measure time? The Earth rotates on its axis, which causes the Sun to appear where it was during the Earth's previous rotation. We call that period one day, which we divide into 24 hours, hours into minutes, and minutes into seconds. But if the Earth did not rotate, the stars were motionless, and if the Moon were stationary, time would have no meaning. Space, without motion, would make the passage of time meaningless. Similarly, if all things in the universe moved *absolutely* with respect to one another, they would all be motionless compared to one another. Therefore, Leibniz refused to accept that space and time were absolute.

So it was clear that Newtonian mechanics and his theories on gravity were not perfect. However, this in no way detracts from Newton's contributions to science. Even during his lifetime, his peers readily acknowledged him as one of the greatest scientific minds. Newton's work stands above most because he did something that no other before him could; he made the motion of the heavens fathomable.

For thousands of years, humans looked at the Moon. They watched it float effortlessly across the sky. It was mythical, inexplicable. Newtonian mechanics changed this view, bringing an end to the *old-world* thinking shaped by Aristotle. Isaac Newton single-handedly ushered humanity into a new age of scientific innovation and the Industrial Revolution. Even his methods remain at the core of how research is still conducted today. His laws of motion, theories on gravity, and calculus are still in use at every level of engineering and science. Hence, many of the last 400 years of scientific discoveries can be traced back to Isaac Newton in theory, mathematics, or methodology. And yet, as large a leap forward as Newtonian science was, something was missing from his works. Newtonian mechanics explained the universe's motion, but it didn't explain how it was *fueled*. It was now time for electricity and magnetism.

Electricity and Magnetism

While Newton dominated the science of mechanics, there are two areas that he remained noticeably ambiguous about: electricity and magnetism. He mentions both in his writings, indicating he did conduct important experiments with these. Yet, it appears that Newton could never fully grasp these phenomena as firmly as he did gravity. Therefore, he never made any breakthroughs when studying electricity and magnetism as he did in mechanics. Still, his notable achievements forever excuse his lack of insight regarding the latter. Besides, as Newton had done for mechanics, the English physician and physicist William Gilbert (1544-1603) had started with electricity and magnetism.

William Gilbert (1544-1603)

In 1603, Gilbert published his book *De Magnete*. The book proposed many ideas about electricity and magnetism that would later be proven correct. For example, he was the first to suggest that the center of the Earth was magnetic and that this was why a compass always points north. The alternate view of the day was that a large magnetic island was located in the north, and compasses pointed toward it. Or that compasses pointed toward Polaris, the pole star. It was also Gilbert who gave us the name: Electricity. In the 17th century, electricity was commonly produced by rubbing a rod composed of amber against wool. While examining this static, Gilbert called the sparks *electric force*, taken from the Greek word for amber – *elektron*.

Consequently, by the time of Isaac Newton's death in 1727, most in the scientific community had come to believe that humanity had reached the apex in its understanding of the physical universe. It was thought that there were only three universal forces: gravity, electricity, and magnetism. It was also believed that these forces were conveyed through space via the aether as different frequencies of pressure waves. And even though Gilbert's explanation of electricity and magnetism was, at best, rudimentary, scientists were confident that they would figure these out as well. And so, for several decades, Gilbert's explanation of electricity and magnetism remained the definitive work. It was not until 1820 that the next significant breakthrough was finally made, and it was entirely by accident.

Hans Christian Oersted was a Danish physicist, chemist, and science professor at Copenhagen University. He was also a very competent clinician who made several notable discoveries. He was the first to develop a method of isolating the compound known as *piperine*. The latter is the alkaloid responsible for the sensations of heat and pain in black pepper. In 1825, he was also the first to isolate the element aluminum. Oersted would become Secretary of the Royal Society of Sciences in Copenhagen, a Knight of the Prussian

Hans Christian Oersted (1777-1851)

Order of Merit, and a French Legion of Honor member. However, he is not primarily remembered for these honors. The most well-known discovery that Oersted made, for which he had a physical law named after him, occurred in April of 1820 while he delivered a lecture to friends and students.

As Oersted applied a current to a wire, he, by chance, noticed that a nearby compass immediately deflected, then returned to its initial position. Despite his immediate excitement, he is said to have merely completed his demonstration as calmly as he could. It took another three months of repeated experiments for him to feel confident enough to publish his findings. And although he could only state his experimental results without explaining why the deflection occurred, the implications were clear: there was an undeniable link between electricity and magnetism. From this point in history onward, electricity was no longer spoken of without mentioning magnetism; the field of *electromagnetism* had been founded. Even so, though Oersted had made a considerable breakthrough, there was yet another phenomenon of nature that continued to elude scientific investigation: light.

"Let There Be Light."

Almost 20 years after completing his work *Philosophiæ Naturalis Principia Mathematica*, Newton published his book *Opticks*. In this book, he described his experiments with light using prisms. It is from these experiments that Newton made some very insightful deductions. For instance, the Aristotelian belief was that white light from the Sun was unadulterated and pure. By passing that light through a prism, however, Newton demonstrated that this light was composed of the colors of the rainbow: red, orange, yellow, green, blue, indigo, and violet. This process, the separation of light into its constituent colors, is called *refraction*. Newton believed this occurred because each color that makes up the ray of light is refracted (or bent) to a different angle after passing through the prism. (Figure 3-5) His hypothesis on refraction was correct. And they proved to be of immediate help to

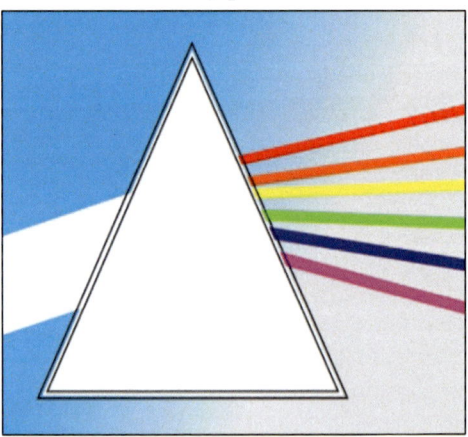

Figure 3-5. White light is composed of all the colors found in the rainbow, and each color has a different wavelength. When white light is passed through a prism (or a lens), each wavelength refracts (or bends) to a different angle upon exiting the prism.

astronomers of his day who were attempting to resolve one of the leading problems with 17th-century telescopes.

To see farther out into space, astronomers of the 17th and 18th centuries created longer telescopes, which contained wider and *thicker* lenses. One such telescope, constructed by the Polish-Lithuanian mayor and amateur astronomer Johannes Hevelius, measured 46 meters long. (Figure 3-6) But an inexplicable

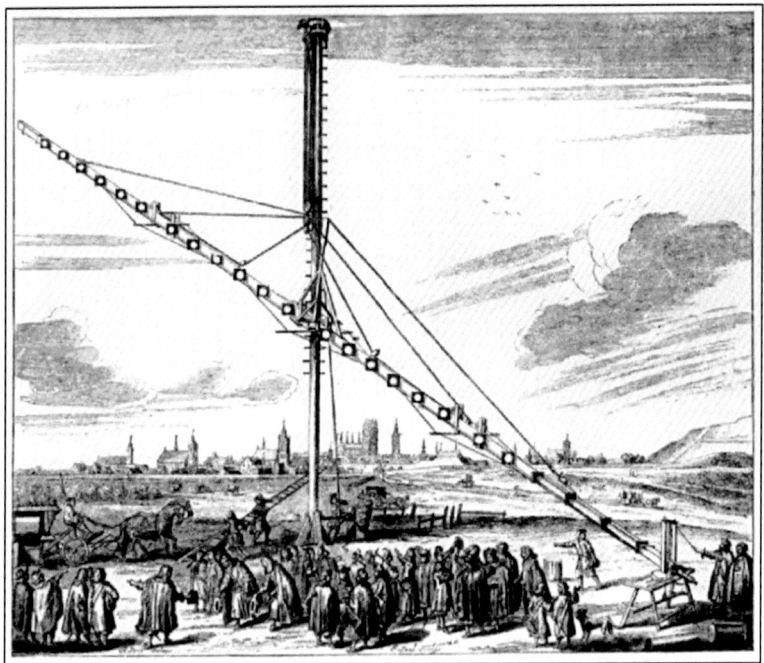

Figure 3-6. The 46-meter telescope built by Johannes Hevelius.

problem had arisen. As the lenses became thicker and the scope longer, the more the light from the object being viewed began to distort into a spectrum of colors. The result was that objects being viewed through these more powerful telescopes were highly distorted. The different colors of light that made up the image were being refracted to ever greater angles as they passed through the thicker lenses. Before Newton's *Opticks*, there appeared to be no way to compensate for this distortion. Yet, Newton did not just explain the reason for the distortion but, going further, also suggested a solution. Instead of making lenses thicker and

99

telescopes longer, Newton constructed the first *reflecting* telescope. (Figure 3-7)

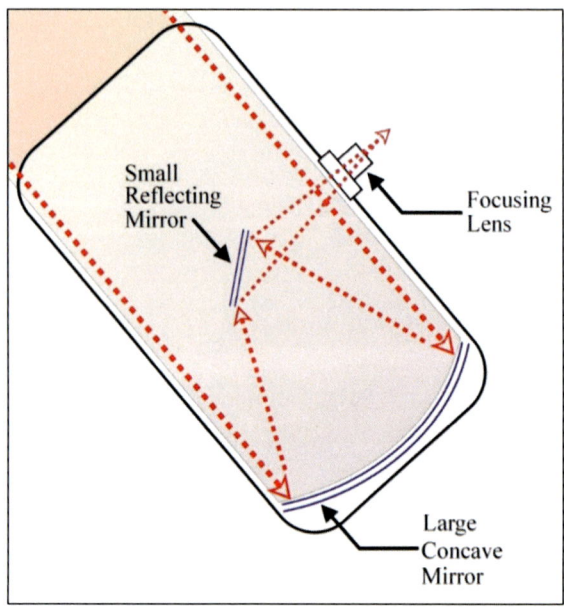

Figure 3-7. A reflecting telescope works by 1) gathering a large amount of light through a large opening, then 2) using a concave mirror to focus that light onto a smaller, secondary mirror, which, 3) reflects that light into an eyepiece for the astronomer to make observation through.

A reflecting telescope works by gathering light onto a relatively *wide* concave mirror at the back of the scope. The light is then *reflected* and *focused* onto a second mirror, which then sends the light to an eyepiece.* This improvement removed all the distortion caused by passing the light through a thick magnifying lens. Still, although well received by the scientific community in general, Newton's book did have its critics.

---

* The person to first suggest a reflecting telescope was the Scottish astronomer and mathematician James Gregory in 1663. However, because the focal point of a reflecting telescope resided inside the scope, Gregory saw no way of constructing it.

One of Newton's most divisive claims about light was his suggestion that it was composed of tiny *particles* called corpuscles. This conclusion put Newton at odds with Robert Hooke and Christiaan Huygens; the latter is chiefly remembered for proving that light was a wave, not a particle. Over time, the debate between three of the 17th-century foremost scientists became so heated that the stress caused Newton to abandon his research in the field of optics completely. He even delayed the publication of his book

Christiaan Huygens (1629-1695)

Opticks until after Hooke's death in 1703. But most ironically, in this last book, Newton was forced to use some of Hooke's wave theory observations to explain the behavior of light.[*] But in the end, the key to understanding the nature of light would only be made after someone could successfully measure its speed, which was no easy task. To understand the difficulty in doing so, let's consider Galileo's primitive attempt to measure the speed of light.

Galileo's attempt to measure the speed of light began by having his assistant stand some distance away; both men held covered lamps. Galileo then uncovered his lamp. Next, using his pulse as the measure of seconds, he immediately began counting the time it took for his assistant to respond by removing the hood from the lamp that he held. In 1638, Galileo published his findings. He stated that the speed of light was at least ten times faster than that of sound (or 3.4 km per second). As you might suspect, this method of measuring the speed of light did not provide accurate results. However, the next person to measure the speed of light was the Danish astronomer Ole Rømer. Being sounder than Galileo's, Rømer's method would get him to within 33% of light's actual speed.

---

[*] Einstein would show that both men were correct. We now know that light has a dual nature: At times, it behaves as a particle, and at other times, like a wave. So today, we're accustomed to speaking of light waves when light behaves as a wave but then calling it a photon when it acts as a particle.

Ole Rømer
(1644-1710)

Rømer was observing the moons of Jupiter through a telescope one evening and noticed something very peculiar. As he estimated when Jupiter would appear over several months, his calculations fell further and further behind. First, by seconds, and then by minutes until, just as inexplicably, the time lag began to reverse. When he pondered the reason for this persistent delay, he realized the answer. As the Earth orbited the Sun and its orbit brought it near Jupiter, the timing for Jupiter's appearance was precise. But when the Earth was at the opposite side of its orbit, the time lag peaked. Rømer had correctly surmised that the delay was caused by the extra time it took for light from Jupiter to reach the Earth when the Earth was on the far end of its orbit. (Figure 3-8) Rømer then estimated the diameter of Earth's orbit, divided that by the delay, and calculated that the speed of light was approximately 200 000 km per second. His estimate was the first to even approach the actual speed at which light traveled. His error was in estimating the time it took for the light to cross Earth's orbital diameter. He believed it was 22 minutes; it was actually only 17 minutes. However, while Rømer's method was sound and his results were close, in the end, the person credited with correctly determining the speed of light was the English physicist and astronomer James Bradley.

James Bradley
(1693-1762)

In 1725, James Bradley used a friend's rooftop as an observatory to repeat one of Robert Hooke's experiments. Bradley was trying to measure the parallax of a star known as Gamma Draconis. As he observed the star's position for over a year, he noticed that he could not detect any parallax. Today we are aware that Gamma Draconis is far too distant for Bradley's instruments to have seen any parallax. However, that year of observations was not wasted; Bradley had made another, unexpected discovery. To keep

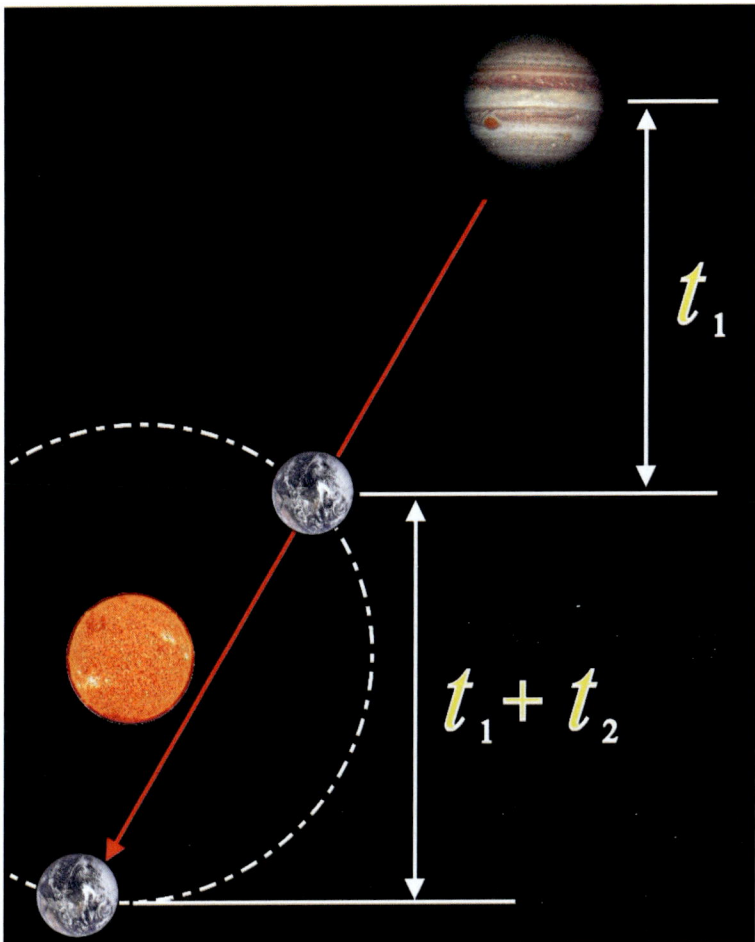

Figure 3-8. Rømer realized that the delay in his viewing of Jupiter was due to the Earth's changing position. When the Earth's orbit brought it closer to Jupiter, the light reached the Earth sooner (t_1). However, when the Earth was at its farthest point from Jupiter, the light took an extra 17 minutes to reach the Earth ($t_1 + t_2$).

Gamma Draconis in view, he was forced to make minor angular adjustments to his telescope each night. He had discovered what we today call *stellar aberration*. The easiest way to understand this effect is to consider two persons caught in the rain.

The first person stands in the rain. If there's no wind, the best angle for him to hold his umbrella so as not to get wet is straight up. A second person is in a hurry, so he's running. What's the best angle for the second person to hold his umbrella to block the most rain? Since there is no wind, the rain is still falling straight down as in the first instance. However, because he's running, the raindrops *appear* (relative to him) to now have a horizontal motion. These two motions (the falling rain and his running) make it appear that the rain is falling at an angle. Therefore, to block the most rain, the second person must tilt his umbrella forward in the direction he's running. This same principle applies to light arriving on the Earth from space.

Since the Earth is in motion around the Sun, the light that arrives from space *appears* to have an added horizontal motion. This motion is equal to but in the opposite direction of the Earth's orbit. Thus, just like the umbrella in our example above, to get the best view of certain stars, telescopes must be tilted forward just a bit in the direction of the Earth's travel. Knowing the angle that he tilted his telescope forward and using trigonometry, Bradley was able to calculate the speed of light. He determined that the speed of light was 10,210 times faster than the Earth's velocity. (Figure 3-9) This estimate put the speed of light at 304 016 km per second, making Bradley's calculations less than 1% off from its

Figure 3-9 (left). If the Earth were motionless, light from each star would arrive at their particular angle of *a*. However, because the Earth orbits the Sun, **Figure 3-9 (right)**, the light from stars instead arrive at an angle of *a'*. The angle *appears* different for the same reasons that raindrops appear to fall at an angle when a person is running through the rain. In astronomy, this effect is called stellar aberration.

accepted value today.* And once this value for the speed of light had been sufficiently determined, the debate now shifted. Now everyone began asking: What exactly is light?

By the late 17th century, the scientific community was split into two opposing camps regarding light's nature, and two of history's greatest minds led either side of the debate. On the one side were those who believed Newton, light was composed of discrete particles. While the others sided with Hooke, that light was a wave. The arguments between the two men became intense.† So, trying to end the dispute, in a manuscript presented to the Royal Society of London in 1675, Newton stated:

They, that will, may suppose it an aggregate... Others may suppose it multitudes of unimaginable small and swift corpuscles... To avoid dispute, and make this hypothesis general, let every man here take his fancy; only whatever light be, I suppose it consists of rays differing from one another in contingent circumstances, as bigness, form, or vigour."

Royal Society, Dec. 9, 1675 –

In addition to the 17th-century controversy regarding the nature of light was the related belief in the aether. Fortunately, it would be an experiment designed to detect the latter that would settle the debate about the nature of light. However, this experiment would not come for another 200 years after Newton's address to the Royal Society. In the meantime, we must next examine the pioneering work of a certain mathematician. We do so because the solutions to four unrelated formulas changed how light and electromagnetism would forever be viewed. Even more importantly, these four equations formed the overall foundation of Einstein's special theory of relativity. However, at the time, the

---

* The accepted value today is 10,066. Meaning that the speed of light is 10,066 times faster than the velocity of the Earth's orbit around the Sun.

† Newton and Hooke did not just argue over the nature of light; they also argued over who came up with the idea for universal gravitation. Hooke stated that he was the one who gave Newton the initial idea, so he wanted equal credit for its discovery. Newton denied his claims of co-discovery.

solution to these formulas seemed so outrageous that the mathematician who discovered them had difficulty accepting them.

Chapter 4

Between Isaac Newton and our next contributor to the theory of relativity, approximately 200 years passed. And over this period, many scientific discoveries were made, especially in regard to electromagnetism and chemistry. For instance:

- Benjamin Franklin confirmed that lightning was a form of electricity (1751).

- Antoine Lavoisier discovered the law of conservation of mass (1789). Later, when coupled with John Dalton's atomic theory (1805), they became the foundation of modern chemistry.

- Michael Faraday discovered electromagnetic induction (1831).

These and other notable breakthroughs in the intervening years kept interest in physics high and science progressing. However, in the same year that Faraday discovered electromagnetic induction, the next contributor to relativity was born: James Clerk Maxwell.

James Clerk Maxwell (1831-1879)

Early in the 19th century, a prospering lawyer named John Clerk stood to inherit a modest parcel of land in Middlebie, Scotland. However, to maintain the family name, it was stipulated that only an heir still having the name "Maxwell" could receive the property. Subsequently, John added the surname Maxwell behind his own. Later, in 1826, John Clerk Maxwell married Frances Cay, the sister of a close friend. An uncommon occurrence for the era was that, upon marriage, both John and

Francis were well into their thirties.* Their first child, Elizabeth, died in infancy. So, James, born when Frances was 40 years old, would be the couple's only child.

James Maxwell was very close to his mother; he even grew to share many of her interests. Aside from being a deeply religious woman, she was also very well-read, thus the reason she decided to home school her son. As a result, though spending much of his early years secluded, Maxwell flourished intellectually. Indeed, along with his natural genius, his eidetic memory was evident even as a child. He is said to have been able to recite complete passages of the English poet John Milton, as well as the entire 119th Psalm; the latter passage is the longest chapter in the Bible, some 176 verses. However, in 1839, at only 48 years of age, Maxwell's mother died of stomach cancer.† As can be imagined, her death had a tremendous effect on the family. And although Maxwell's father endeavored to continue to nourish his son's intellect, he also became incredibly protective of his only child.

At first, Maxwell's father hired a tutor to continue his son's education at home. But, from extant writings, the tutor is described as a "bully who struck the child often." So Maxwell's father decided to send his son to public school. Yet, this too proved to be a challenge. Having been homeschooled from infancy, Maxwell was relentlessly teased because of his strong accent and country dress. Thus, by the time he entered high school, he was reported to have been highly introverted, more prone to reading alone than socializing. Still, despite all of this, young Maxwell remained intellectually curious. By the time he was 14, he had already written his first paper. It was on geometry and described a more simplified method for drawing curves. And although its content was not groundbreaking, it was nonetheless presented to the Royal Society of Edinburgh. Then, in 1847, at just 16 years of age, Maxwell was accepted to the University of Edinburgh. By the time he was 18, he had authored two more papers. These papers, too, were decidedly mathematical, a testament as to where his talents were leading him. And yet, like many we

---

* On average, in 19th-century England, men married by the time they were 28 and women by 22 years of age.

† Sadly, this same illness took Maxwell's life at the same age.

have thus far considered, Maxwell's scientific contributions went far beyond mathematics.

Maxwell's work in optics provided the basis for modern color photography. He was the first to produce a practical paper on control theory – the branch of engineering and mathematics concerned with the designing, constructing, and the analysis of interactions between components in complex mechanical systems. Even more than his contributions to relativity, Maxwell's work with gases is still an essential part of thermodynamics; his formulas in this latter discipline are still referred to as the *Maxwell Distribution*. And yet, while all these contributions are worthy of discussion in-and-of themselves, the breakthrough that Maxwell is most famous for (and the one we're most interested in) is in regard to his research into electromagnetism. And even though he was not the first to derive the following equations, he was the first to discover how they interrelated. Because of this achievement, these four equations, as a group, would forever bear his name.

Maxwell's Equations

Although Maxwell is considered among history's most prominent theoretical physicists – and, indeed, he did study physics while at university – his university degree was in mathematics. For this reason, his contributions toward the theory of relativity were especially mathematical, even more so than those of Newton. Therefore, to fully expound upon Maxwell's equations would require the inclusion of more mathematics than promised. Hence, while we list the equations, we will not derive them. Instead, we'll merely explain what each equation tells and, most importantly, the consequences that must necessarily follow. As shall be seen, Maxwell's genius lay in the fact that he saw the relationship that existed between four seemingly unrelated formulas. It was a relationship that would reveal a universal truth so bizarre that even Maxwell had difficulty accepting it.

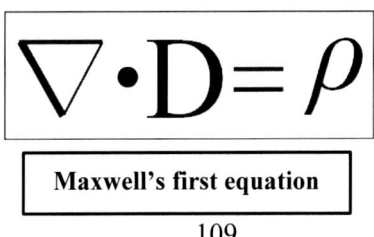

$$\nabla \cdot D = \rho$$

Maxwell's first equation

Maxwell's first equation was initially formulated by the man still considered Germany's greatest mathematician, Carl Friedrich Gauss; it's called Gauss' Law for electricity. This equation relates the strength of an electric field radiating from an object to the intensity of that object's electric charge. So, for example, the smallest *negatively* charged particle is the electron. And, because the electron has a negative electrical charge, it generates a negative electrical field. Furthermore, when an electron is *not* in motion, the electric field it radiates is symmetrical. This first equation is important because it tells us that:

1. The electric field lines of a positively charged object will diverge *away* from the object.

2. The electric field lines of a negatively charged object will diverge *toward* the object.

The effect of imagining positive and negative electric fields as diverging in different directions is that we can conceive them as always attracting one another: opposites attract. Therefore, concerning electromagnetic behavior, Maxwell's first equations tell us that oppositely charged objects will always attract, while similarly charged objects always repel. It's as simple as that.

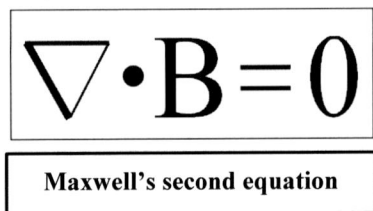

$$\nabla \cdot B = 0$$

Maxwell's second equation

Gauss also discovered Maxwell's second equation or Gauss' Law for magnetism. In this second equation, the essential thing to note is that, although it's very similar to the first, the answer it yields on the right is zero. From this, we conclude that a magnetic field's strength *does not* relate to the intensity of the magnetic charge. Why is this true? Because there's no such thing as (at least no one has yet discovered) a magnetic *charge*. So, what exactly is this equation telling us?

Whenever we think of an electric charge, we intuitively know that it can be either positive or negative. For example, a positive proton can exist without an electron nearby to balance it, or vice versa. But Maxwell's second equation relates to magnetism, and there's no such thing as a magnetic *charge*. Instead, magnets *always* have two poles, and they <u>do not</u> exist apart from one another. Even if you cut a magnet in half, new north and south poles are created. And that's why the answer to Maxwell's second equation is *always* zero; a magnet will *always* have a north and a south pole present, and these *always* cancel one another.

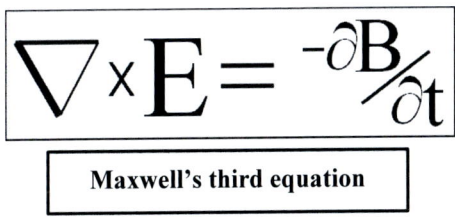

Maxwell's third equation

Maxwell's third equation links the first two; they are related by *voltage*. This equation was discovered by Michael Faraday and is more often called the Law of Induction. To understand this third equation, consider Figure 4-1.

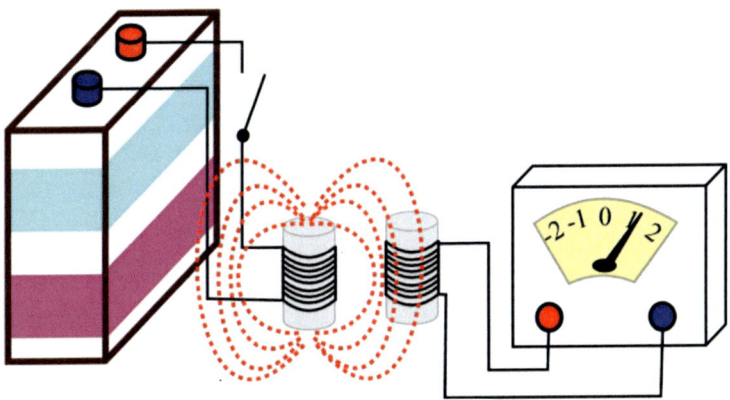

Figure 4-1. A magnetic field is generated when electricity flows from the battery and through the wire wrapped around the first iron core. When the magnetic field from the first iron core encounters the second iron core with a wire wrapped around it, an electric field is produced in the second iron core. Additionally, as registered by the ammeter, a current is also briefly produced in the second wire. All this occurs even though the iron cores are not in direct physical contact with one another.

Imagine a battery with a wire connected to its positive terminal. The wire is then wrapped around an iron core before returning to the battery's negative terminal. Next, a short distance away from the first iron core is a second iron core with a wire wrapped around it. However, instead of being connected to a battery, this second wire is connected to an ammeter. When the battery is switched on, even though the two coils are not in direct contact, the ammeter will still register a current flow before returning to zero; this is magnetic induction. It occurs when the <u>flow</u> of electric current through the first coil creates a magnetic field (i.e., equation 2). Then that magnetic field *induces* a voltage in the second coil (i.e., equation 1).

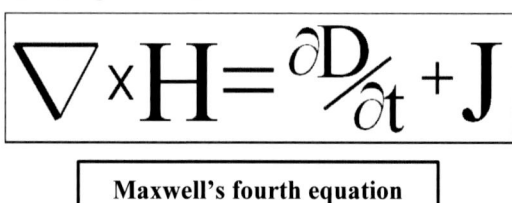

$$\nabla \times H = \frac{\partial D}{\partial t} + J$$

Maxwell's fourth equation

Finally, Maxwell's fourth equation was discovered by the French physicist and mathematician André-Marie Ampère. In honor of his discovery, we call the flow of electrons amperage (or amps for short); Ampère discovered Ampère's Law. Simply put, this equation reveals that, just as a magnetic field can produce an electric field, an electric field can produce a magnetic field. And these are the four equations that describe the behavior of electromagnetism. Indeed, these equations form the entire basis of the study of classical electrodynamics. However, regarding the theory of relativity, the final two of Maxwell's Equations proved to be the most enlightening.

So, according to Maxwell's third equation, an alternating magnetic field produces an electric field. And according to the fourth equation, this changing electric field creates a magnetic field, which in turn produces a new electric field; *ad infinitum.* Yet, this indefinite creation of magnetic and electric fields went utterly contrary to Newtonian physics. According to Newton, all energies eventually dissipate. But here was a form of electromagnetic energy that, once created, could self-propagate by indefinitely alternating between an electric, then magnetic waveform. And yet, this was not even the most surprising aspect of Maxwell's discovery.

In 1865 Maxwell published his work *A Dynamical Theory of the Electromagnetic Field*. After showing that this alternating electric/magnetic field <u>did not</u> exist in the wires, but it traveled through space as a wave, he next mathematically derived its speed: 300 000 km per second. It was the speed of light! Hence, as Oersted had combined electricity and magnetism some 45 years prior, Maxwell had successfully unified the study of electromagnetism and light. Light was an electromagnetic wave; it was an alternating electric-magnetic field that self-propagated through space. (Figure 4-2a)

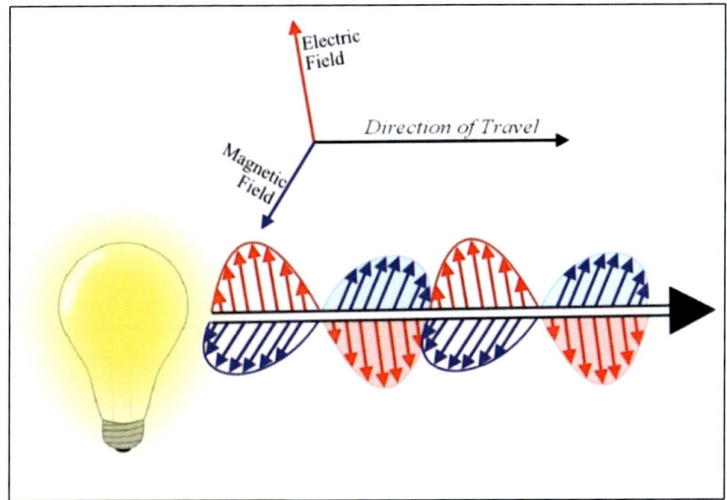

Figure 4-2a. Maxwell's equations describe the behavior of electromagnetic (EM) waves. As EM waves propagate through space, each opposing field produces the next. Thus, the first electric field produces the next magnetic field and vice versa. Furthermore, all EM waves travel at the speed of light: 300 000 km per second.

It was now the late 19th century, and physicists again thought they were on the verge of understanding everything. Within 200 years of Newton's successes with his laws of motion, they had reduced the number of universal forces from three (gravity, electricity, and magnetism) to two (gravity and electromagnetism). Yet, this was not Maxwell's only breakthrough. While arranging his equations, Maxwell discovered something even more peculiar. Every time the speed of light appeared in his calculations, it *always* occurred as a constant. It always had a constant speed of 300 000 km per second.

Now, just as with those before him, Maxwell believed in the aether. He described it as an all-pervasive "sea of molecular vortices" that carried the electromagnetic wave, the same as air transmitted sound. However, belief in both an aether and the speed of light as a constant is difficult to reconcile. For example, suppose the aether did transmit electromagnetic waves; that would mean that its transmission must occur due to vibration. One aether *particle* causes the next to vibrate as the wave propagates. But for this to happen, since light has an extremely short wavelength, the aether would have to be dense, as dense as steel! Yet, given that the Earth, Moon, and every other material object in the universe can easily move through space, a dense aether is implausible. Also, if the aether did exist, it would actually disprove Maxwell's calculations that the speed of light was constant. To understand why the latter statement must be true, let's imagine a steamboat traveling up a slow-flowing river. (Figure 4-2b[*])

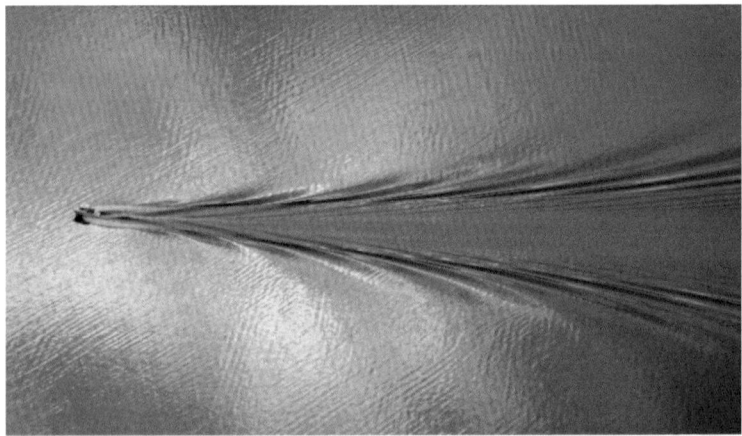

Figure 4-2b. Just as the waves are compressed at the boat's bow and stretched at its stern, if the aether existed, the same would be true of light as it traveled through it.

[*] Image credit:
https://commons.wikimedia.org/wiki/File:Fjordn_surface_wave_boat.jpg

Water waves propagate through vibration; each water molecule transmits its motion to the molecule besides it. But what happens when waves attempt to travel against the current? If the river is flowing slowly, the waves might travel a slight distance upriver. However, they will never go that far. Even a slow-moving current compresses and dissipates waves. Meanwhile, the waves that form at the back of the boat are stretched; this occurs because, as the boat moves forward, the river's current flows away from the boat's stern, stretching and elongating the waves. (Figure 4-2b) Indeed, this example is only valid if the river's current moves relatively slowly. If it's a fast-flowing river, no waves whatsoever will travel upstream. Similarly, if the aether existed, the same principles would apply to light waves.

Imagine a light-emitting object like the Sun existing in a flowing aether. For the same reason as the waves that form at the boat's bow in the previous example, light emitted from the front of the Sun would also be compressed and would quickly dissipate. On the other hand, the light emitted from behind the Sun would be stretched and elongated. And if the aether flowed at a right angle to the Sun, it would be like a riverboat traveling across a river. In this latter case, the light the aether transmitted would be swept "downstream," and the star would not be seen in any other direction. But what if we suppose that matter didn't interact with the aether and that the aether didn't flow? Would this resolve the problem? No, it wouldn't.

We know that light travels at 300 000 km/second. So if the aether existed, it must transmit light at that constant speed. Now, suppose the matter that made up the Sun and the aether did not interact. Then, as sunlight exited the Sun and entered the aether, it would be accelerated to a constant speed of 300 000 km/second; the speed of light would be constant. Yet, this creates a new problem since we know that the Sun is not static.

As large as it might be, the Sun rotates around the center of the Milky Way Galaxy. Therefore, in a static aether, the sunlight emitted in the Sun's direction of travel will now have the Sun's added momentum as it (the light) enters the aether. It's the same reason that bullets fired from the front of a fighter jet can reach the target jet, even at supersonic speeds. The velocity of the firing jet adds to the speed of the bullet. Similarly, the velocity of the Sun traveling through a static aether would add to the sunlight's speed. Thus, the speed of light would not be constant. By the same logic, light emitted opposite the Sun's travel direction would be moving slower than light speed. Therefore, just as with a flowing aether, assuming that matter does not interact with the aether still results in the speed of light <u>not</u> being constant.

Of course, another way to resolve the problem is to argue that the speed of light is constant in relation to the aether. Thus, no matter the momentum of the Sun, the aether can only transmit light at the speed of light. The problem with this solution is that you would need to reject the idea of Galilean relativity and return to the notion of space being absolute. You would merely be replacing the Earth with the aether as the absolute frame of reference from which to measure all speed. Thus, Maxwell's equations had only made it clear that the existence of an omnipresent aether was impossible to reconcile with the speed of light being constant.

As a result, Maxwell's equations were a double-edged sword; they revolutionized our understanding of light but also introduced seemingly *logical* contradictions. Yet the equations were correct. And to further complicate the issue, a vocal minority of scientists argued that the aether had to be real, despite Maxwell's Equations being accurate. Since the ancient Greeks, humanity had known that waves require a medium to propagate through. And this knowledge was backed by a good deal of evidence; in other words, the aether had to exist. Consider sound waves.

Sound waves occur when an object in the air is vibrated. As these vibrations compress and decompress the air, the sound is propagated in the form of a pressure wave. Confirming that a medium (air) was required to transmit sound, physicist Robert Boyle repeated an experiment first conducted by Jesuit scholar Athanasius Kircher. He placed a bell inside a glass jar, and as he continued to ring the bell, he slowly evacuated the air. Boyle listened carefully as the intensity of the sound began to diminish. And the sound only lasted until sufficient air had been removed. At that point, though the bell could be seen ringing inside the jar, no sound was heard. This experiment confirmed that all waves require a medium to be transmitted. Consequently, scientists of the 19th century were confident that Maxwell's Equations showing that light was a wave concurrently proved the aether's existence.

Thus, the next stage in the development of the theory of relativity became obvious: to somehow detect the aether existed. Not much time would have to pass before an experiment was devised.

Edward Morley and Albert Michelson

Early in the 19th century, women in the United States gained the right to attend college. Anna Clarissa Treat was among the first women to enroll in university at the newly founded Hartford Female Seminary in Connecticut. Although these institutions for women were novel, Hartford's curriculum was as rigorous as any educational institution for men. Subjects taught included courses in mathematics, geology, chemistry, and

Edward Morley (1838 -1923)

physics. Later, Anna Treat married the Reverend Sardis Brewster Morley. And when her eldest son Edward was born on January 29, 1838, Anna's education made her more than qualified to home-school Edward until he left for college in 1857.

After Edward Morley earned his Bachelor of Arts degree, he enrolled in graduate school at the Andover Theological Seminary. In that same year, the US civil war began. Within months, most of Edward's fellow students and professors had voluntarily joined the military. John, the first of Edward's brothers, was drafted. And later, his youngest brother Frank volunteered. Anna Morley was distraught. Although readily admitting the evils of slavery, the prospect of losing all three of her sons was more than she could bear. So, to keep Edward from being drafted, Anna convinced her husband to pay the $300 (US) exemption fee, equivalent to $10,000 (US) today. However, she didn't anticipate that Edward would not be able to resist the call to arms. He joined the military in 1863. However, things didn't turn out exactly as he had hoped.

Edward Morley had extensive education in physical science. But since his undergraduate and graduate degrees did not reflect this, he couldn't serve as a military technical officer as he had hoped. Morley was instead made an officer in the newly formed

117

United States Christian Commission (USCC). In effect, he served as a health and social services professional for wounded soldiers. Still, although unable to gain the real-world technical experience he desired, Morley's time in the USCC was not wasted. While serving, he learned the importance of punctuality and exactness, qualities that would make him a very deliberate technical clinician.

When the war ended in 1865, Morley began teaching at a school for boys in Massachusetts. Due to geographical research conducted while in college, he had already gained a reputation as a competent scientist. As a result, Morley was constantly receiving offers to teach at more prestigious schools, offers that, for various reasons, he declined. But, in 1868, he was asked to become the preacher in a small country parish in Ohio, an offer he eagerly accepted. Morley was now in the perfect location to take a position as a chemistry professor at the prestigious Western Reserve College in Hudson, Ohio. It was a turning point in his career. Because while Edward Morley is primarily remembered for conducting one of the greatest "failed" experiments in history, his time at Western Reserve also allowed him to achieve many other notable milestones.[*]

Albert Abraham Michelson was born in what is now Poland in 1852. Soon after his birth, his family immigrated to the United States. During the gold rush, Michelson was reared in Murphy's

Albert Michelson (1852-1931)

Camp, California, then Virginia City, Nevada, during the silver rush. Desiring their son to receive a good education, Michelson's parents sent him to attend high school in the major metropolitan city of San Francisco, California. By all accounts, it appears he did exceptionally well. However, once he completed high school, Michelson found his options for college limited; until he saw an ad in a local newspaper in 1869.

---

[*] Edward Morley was the first person to determine the atomic weight of elemental oxygen and hydrogen precisely. He did much to advance the science of metrology, invented several key experimental laboratory devices, and taught courses in chemistry, botany, and geology.

The newspaper announced that there would be a competitive examination for admittance to the US Naval Academy in Maryland. Young Michelson immediately entered the scholastic contest. And though there was only one position available at the academy, the contest ended with Michelson, along with two others, tied for first place. Unfortunately, the final selection was said to have had more to do with political maneuvering than academic merit; Michelson was not chosen. However, this blatant discrimination outraged the community, to a local congressman's discredit. And what occurred next speaks more about Michelson's personality than any other event in his life.

Due to public outcry, the Nevada congressman whose political expedience decided the contest was compelled to write a letter to then-President Grant. He requested that Michelson receive a special appointment. Thus, young Michelson (with the letter in hand) traveled by train – some 4800 km – to Washington, DC. At that time, it was not uncommon for an ordinary citizen to show up at the White House to ask for, and to receive, an audience with the President. Upon his meeting the President and presenting the letter from the congressman, it is said that President Grant listened very patiently to the young man. However, Grant gently broke the bad news once he had finished pleading his case. He could only appoint ten persons for that year's class, and those appointments had already been made. Young Michelson was visibly disappointed as he left the President's office, believing his future was once more uncertain. But his determination had impressed one of the President's guards. On the off chance that one of the appointees failed the entrance exam, the guard urged Michelson to travel to the academy. This Michelson did in June of 1869.

When he arrived at the academy, Michelson went straight to the Commandant of Midshipmen's office. After waiting for three days, he was finally given an interview and was allowed to take the entrance exam. Yet, upon passing the exam, Michelson was once more told that there were no vacancies for that year's class. This final rejection, and the fact that he was almost out of funds, disheartened the young man. So he made his way back to Washington, DC, to board the train returning to California. However, unknown to him, his persistence had paid off. Right before the train was scheduled to depart, a messenger from the

White House boarded and called out his name; for a second time, Michelson was off to meet with the President.

Michelson, again, nervously stood before the President. He learned that his determination and test scores had impressed one of the examiners, Vice Admiral David D. Porter, Superintendent of the Academy. After Michelson had departed the academy, the admiral dispatched a letter to the President. He requested that the appointment quota be ignored and that Michelson be nominated as the 11th appointment. On June 28, 1869, President Ulysses S. Grant presented Albert Abraham Michelson the 11th appointment-at-large to the US Naval Academy in Annapolis, Maryland. Michelson served in the navy for the next 12 years, with his final two years spent doing scientific research in Europe. He became internationally recognized as an expert in optical research and, in 1907, won a Nobel Prize in Physics. Michelson resigned from the navy in 1881 and, in 1883, accepted a position as professor of physics at Western Reserve College, which by then had been relocated to Cleveland, Ohio; this was where he met Edward Morley.

The Michelson-Morley Experiment

By 1887 science had generally accepted that light was an electromagnetic wave that traveled at a constant speed of 300 000 kilometers per second. What wasn't known was how this was possible, as belief in an aether that transmitted light waves was at odds with light having a constant velocity. Therefore, the next great experiment that had to be devised needed to prove if the aether existed. And interestingly, the person to first conceive of such an experiment was James Maxwell, though he saw no practical way to construct and carry the experiment out. You see, the difficulty with any experiment meant to measure the speed of light is trying to do so accurately. But this was what Michelson and Morley eventually figured out. However, before we examine their findings, let's consider the logic behind their experiment. The following is said to have been the reasoning that Michelson later used to explain the experiment to his daughter.

Suppose two swimmers are preparing to race equal distances of 60 meters. The first swimmer will swim 30 meters upriver and then 30 meters back downriver. The second swimmer will swim across the river 30 meters to the opposite shore and then back to the shore from which he started. Here is the question: If the river's current is 0.9 meters per second, and both swimmers swim at equal speeds of 1.5 meters per second, which swimmer will cover the 60 meters fastest?

The first swimmer starts swimming upriver at 1.5 meters per second. However, the river's current is flowing against him at a rate of 0.9 meters per second. Therefore, his speed relative to the shore is only 0.6 meters per second. It, therefore, takes the first swimmer 50 seconds to swim 30 meters upriver. However, instead of subtracting, we add the current's speed as he makes the return trip; the first swimmer travels the downstream portion of his trip in 12.5 seconds. Thus, the *total* time for the first swimmer to travel 30 meters upstream and then 30 meters downstream is 62.5 seconds. The second swimmer now takes to the water. He doesn't have to fight against the current directly because he's swimming perpendicular to it. But that doesn't mean he's not affected by the river's flow.

As the second swimmer traverses the river at 1.5 meters per second, the current pushes him downstream at 0.9 meters per second. Using simple trigonometry, we can calculate that the total distance the second swimmer must travel to reach the opposite shore increases from 30 meters to 38 meters. (Figure 4-3) Therefore, it takes the second swimmer 25 seconds to cross the river. Adding in his return trip gives a total of 50 seconds for the second swimmer to cross the river and return to the other side. Therefore, the second swimmer, traveling perpendicular to the river's current, can swim 60 meters faster than the first swimmer swimming against and then with the river's current. The second swimmer would win the race. This same logic would apply to light traveling in the aether.

If the aether existed, then the light, traveling in the same direction that the Earth orbits the Sun, would be traveling against the aether. Therefore, it would experience the same effects as the first swimmer going against the river's current. While light moving perpendicular to the Earth's orbit would be comparable to the

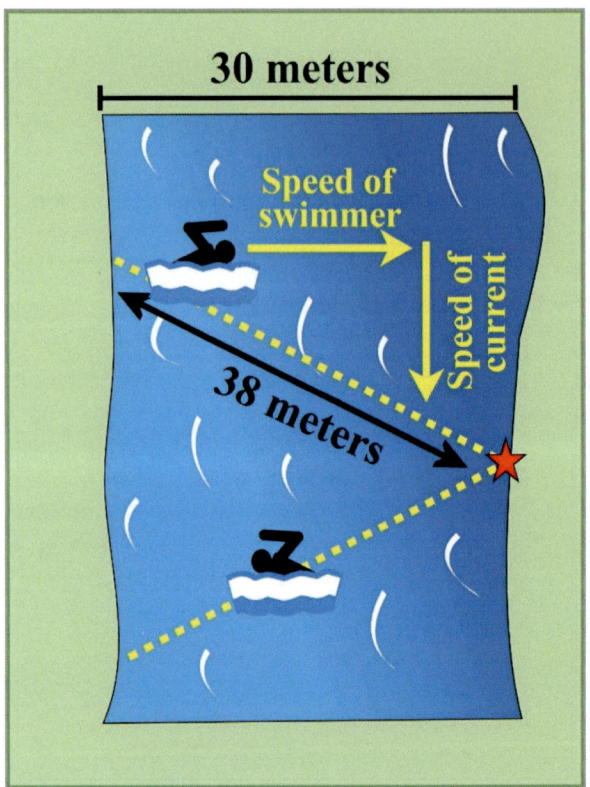

Figure 4-3. The river's current does not allow the second swimmer to travel directly across the river. Instead, as the current pushes him downstream, he is forced to swim a longer distance. Yet, the time it takes for him to travel perpendicular to the current is still less than the time for someone swimming parallel to the river's current.

second swimmer who swam perpendicular to the river's current. But knowing the logic behind the experiment still wasn't enough. They next had to overcome the issue of setting up the experiment, beginning with eliminating all external vibrations.

Since visible light has a very short wavelength, even slight vibrations, such as those caused by a passing horse-drawn carriage, could distort the experiment's results. They began by placing their experiment underground, inside the stone-walled basement of the school dormitory. Next, the apparatus sat atop a thick sandstone slab that floated in a pool of liquid mercury. These precautions were sufficient in eliminating external vibrations. However, resolving this first problem was easy compared to the next one: How does someone in the 19th century accurately measure the speed of light?

This second problem was the real challenge. Michelson and Morley needed to precisely measure – then compare – the speeds of two light beams traveling at right angles to one another. Yet, without computers or modern electronics, they had no device precise enough to measure the speed of light across short distances. So, attempting to measure the speed of light as one would time two athletes in a foot race was out of the question. But all hope was not lost, and this is where the brilliance of Michelson and Morley was manifested. They found the solution to their second problem within Maxwell's equations; in the fact that light is and, therefore, *behaves* as a wave.

Although sound and light differ in how they are produced and propagate, both still possess two properties that all waves share. First, all waves have a wavelength – the distance between two identical points of successive waves. Second, all waves have a frequency – the amount of time between the same point of two consecutive waves. Understanding the above, Michelson and Morley realized that they wouldn't need to measure the two light beams directly. Instead, all they would need to do is split a single beam of light, then send the two beams along equal but perpendicular paths several times. Finally, they would recombine the two beams. If the aether did exist, when the two beams are recombined, the two waves would no longer be in phase (i.e., their frequency and wavelength would not realign). To understand this process, consider the siren of a passing emergency vehicle.

As an emergency vehicle approaches, the sound waves from its siren are compressed; thus, the frequency is shortened. As a result, the siren's sound is resonant and distinct. (Figure 4-4a) However, once the vehicle has passed, the siren begins to take on a duller tone. As the emergency vehicle travels away, the sound waves behind it are stretched, creating a longer frequency as the vehicle speeds away. As a result, even though the same siren produces the sound, the sound waves ahead and behind the vehicle are now at different frequencies. Consequently, if you attempt to recombine these separate sound waves, placing one on top of the other, they would no longer be perfectly aligned. (Figure 4-4b) This same principle would apply to light if the aether existed.

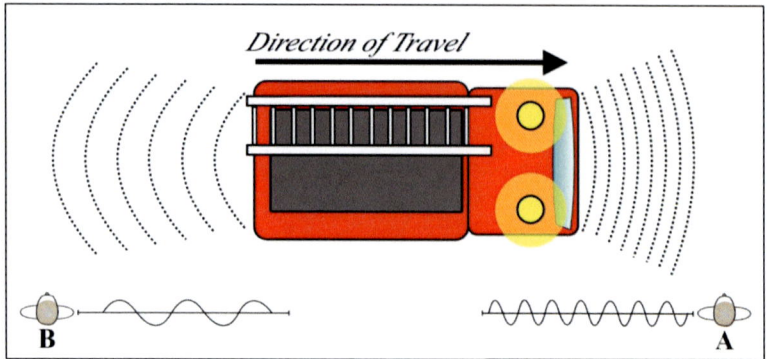

Figure 4-4a. The person standing at position **A** hears the siren more distinctly than the person standing at position **B**. The vehicle's velocity alters the siren's frequency, compressing the sound waves ahead of the vehicle, stretching the sound waves behind it. This change in wavelength is called a Doppler shift.

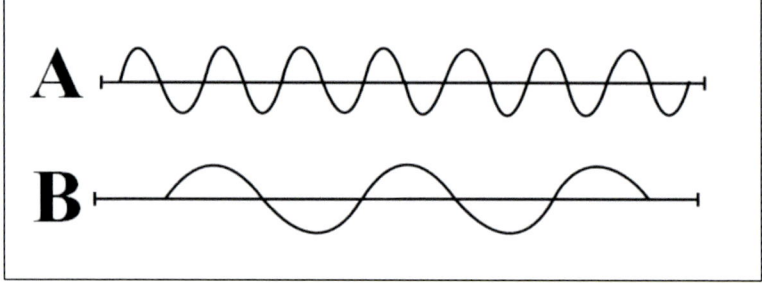

Figure 4-4b. Waveform **A** is the siren's compressed sound wave, which is heard by the person standing ahead of the moving vehicle in Figure 4-4a. Waveform **B** is the elongated sound wave of the siren, heard by the person standing behind the retreating emergency vehicle. As can be seen, although the same siren emits the two waves, the two waveforms no longer have the same frequency.

Light waves emitted from an object traveling at a significant speed through the aether would behave exactly like the emergency vehicle's siren. And if you were to recombine these light waves, their waveforms would no longer align. Instead, they would now interfere with one another, producing a light pattern containing bright lines where their peaks aligned and dark lines where they did not. (Figure 4-4c) The latter would be the telltale sign indicating that the aether *did* exist. So, what happened?

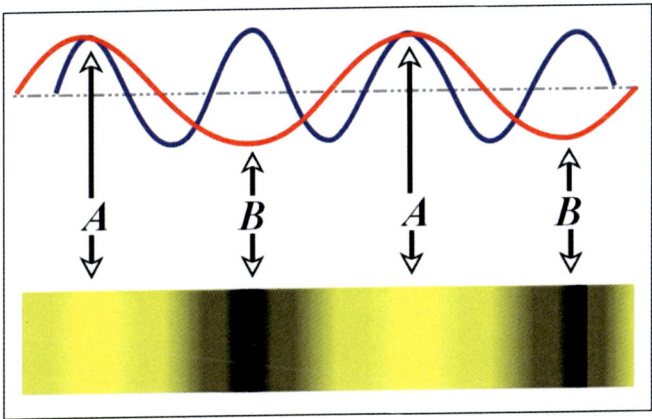

Figure 4-4c. This is an example of the type of diffraction pattern that would have proven to Michelson and Morley that the aether existed. The bright areas (*A*) are where two peaks, one from each light wave, perfectly align and thus amplified one another: *constructive* interference. However, where there's a dark slit (*B*), one light wave's peak overlaps a second light wave's valley. The result is the waves cancel, and no light is seen: *destructive* interference.

The experiment began with a single beam of light. The beam was then split into two light beams having identical frequency and wavelength. The beams were then sent equal distances but in different directions relative to the Earth's motion around the Sun.* Finally, the beams were recombined. If an interference pattern were seen, such as in Figure 4-4c, this would prove that the aether existed. But if the two beams of light remained in phase, producing no interference pattern, this would be evidence that the aether didn't exist. After Michelson and Morley had carefully set up the experiment (Figure 4-5), ran it several times, and at several angles, what were the results? They never saw a diffraction pattern that indicated that the aether existed, and these results greatly disturbed them.

---

* As measured at the Earth's equator, the Earth rotates on its axis at 1720 km per hour. While the orbital speed of the Earth around the Sun is 172 600 km per hour.

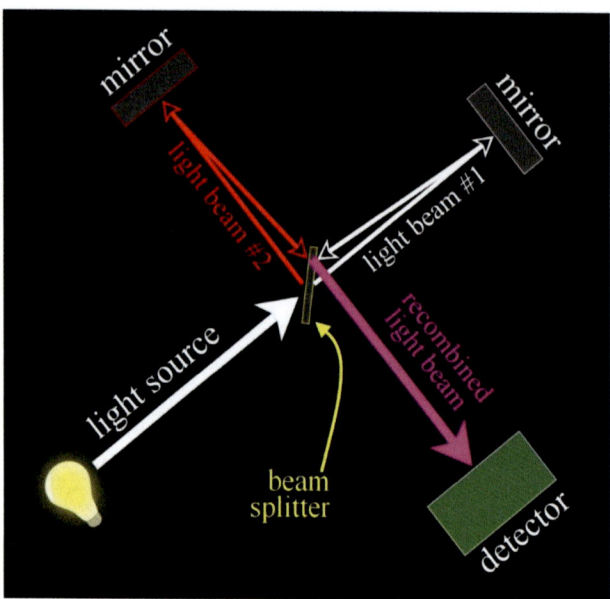

Figure 4-5. The Michelson-Morley experiment sent a single beam of light to a beam splitter. The then two light beams are sent to mirrors, equal distances away but at right angles to one another. Finally, they are reflected back and recombined. The now recombined light beam is sent to a detector to see if they are still in phase. After conducting their experiment several times, Michelson and Morley never found any significant diffraction pattern.

When Michelson and Morley failed to get a diffraction pattern, it never occurred to them they had just proven that the aether didn't exist. Instead, they took it as meaning they had somehow set up the experiment incorrectly, which was how they published their results. And this is why their experiment is regarded as the most successful "failed" experiment in history. For Michelson and Morley, it was easier for them to accept that they had somehow performed the experiment incorrectly than believe that light traveled without the aid of a medium and at a constant speed. Today the Michelson-Morley experiment is considered one of the most ingenious experiments ever conceived. They took a few of the most fundamental scientific principles and then used them to test for the existence of a hypothetical aether. And as other scientists of the day started to review their results, they also found no errors. Accordingly, many had no choice but to accept their results. And finally, as everyone began to explore the consequences of the Michelson-Morley experiment, the path toward the theory of relativity now became outright existential.

FitzGerald, Lorentz, and Poincaré

The failure of the Michelson-Morley experiment to detect the aether immediately ignited a firestorm of discussion. Everyone was attempting to explain why such a well-crafted and executed experiment would not reveal the presence of an aether. The first noteworthy idea would come from an Irish professor named George FitzGerald. Keeping in mind that the accepted belief at the time was that the aether existed, FitzGerald suggested that perhaps the very presence of the aether itself distorted the results. How so?

George FitzGerald (1851-1901)

Begin by imagining a long object moving through a liquid, such as a submarine through the ocean. As the submarine picked up speed, the force of the water against its bow would increase the pressure at the front of the ship. And the faster the submarine moved through the water, the greater the pressure *in the direction of travel*. The consequence would be that this higher pressure would cause the submarine's length to shorten ever so slightly *in the direction of travel*. FitzGerald theorized that this was what happened in the Michelson and Morley experiment.

The increased pressure exerted on the apparatus by the aether *in the direction of Earth's travel* had decreased the apparatus' length. The result of this contraction would be that the light beam that traveled *in the direction* that the Earth orbits the Sun will have actually traveled a shorter distance. Recall, speed is defined as *distance* divided by *time*. Thus, according to FitzGerald, the light beam traveling against the aether had indeed slowed down. However, because the *distance* the light traveled had also been reduced, the *time* it took to travel that shortened distance (to-and-from the beam splitter) remained unchanged. So, the physical *contraction* of the apparatus caused it to *appear* as if the speed of light was the same for both beams.

FitzGerald's proposition – that the length of all objects in the direction of the Earth's travel was *uniformly* contracted because of the pressure placed on it by the aether – was creative. Furthermore, because this contraction affects *all* objects on the Earth's surface, even the instruments used to measure distance, this shortening of length becomes imperceptible. Therefore, according to FitzGerald, the presence of the aether only serves to give the *appearance* that the speed of light was constant.

**Hendrik Lorentz
(1853-1928)**

Meanwhile, and quite independently, Dutch physicist Hendrik Lorentz reached the same conclusion as FitzGerald. Lorentz even devised a formula relating the amount of length contraction being experienced by an object moving through the aether relative to the speed of light. The implication was that the amount of length contraction would *always* be directly proportional to the speed of light. However, Lorentz also realized that length contraction alone could not fully explain the Michelson-Morley experiment. Going one step further, Lorentz also suggested that, along with contraction in the direction of travel, *time dilation* had to be occurring as well. But to understand why length contraction *cannot* happen without time dilation (i.e., the slowing of time), consider the following example of a single photon of light bouncing between two mirrors on a fast-moving train.

Yori is on a bullet train in Tokyo, traveling at 60 meters per second. (Figure 4-6a) In front of him, on either side of the train cabin, there are two mirrors. Yori watches as a single photon of light bounces back and forth between the mirrors for 5 seconds. To Yori, the total *distance* the photon travels is the cabin's width (let's say 2.5 meters across), doing so at the speed of light. And as if he were watching a tennis match, Yori moves his head side to side along with the photon, never taking his eyes off it. Now, Kira is behind Yori on the train. Since the distance traveled by the photon is so small relative to the speed of light, she watches as Yori's head turns back and forth very, very, very quickly.

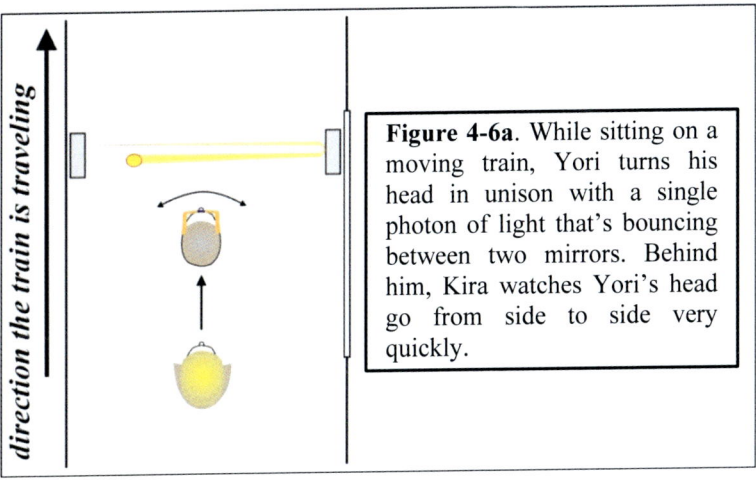

direction the train is traveling

Figure 4-6a. While sitting on a moving train, Yori turns his head in unison with a single photon of light that's bouncing between two mirrors. Behind him, Kira watches Yori's head go from side to side very quickly.

Next, imagine a third person, Sora, who's not on the train. Sora is standing motionless at the station as the train passes. But she, too, can see the same photon for the same 5 seconds that it bounces between the mirrors. Now recall that, as Galileo showed in his ship experiment, the photon's trajectory will appear much differently to Sora than Yori. Sora *does not* simply see the photon merely bouncing the train cabin's width from left to right. Instead, *adding* the train's forward velocity relative to Sora, the photon travels with a zigzag trajectory between the two mirrors. To her, it traverses the 2.5-meter width of the train cabin, *plus* the 300 meters forward during the 5-second time interval (i.e., the train's speed of 60 meters per second, for 5 seconds). (Figure 4-6b) So, during the same 5-second period, the photon has traveled a *greater* distance relative to Sora than Yori. And herein lies the problem.

To Yori, aboard the train, the speed of light is 300 000 km/sec. To Sora, who is motionless relative to the train, the speed of light is also 300 000 km/sec. Yet, also from Sora's perspective, the photon has traveled a greater distance *in the same amount of time*. How can light be traveling *at the same speed* for Yori and Sora yet traverse *different distances* <u>during the same 5-second period</u>? The key to understanding how Sora sees the photon travel a longer distance than Yori, while the photon remains at the same speed for both, is to contrast Sora's view of Yori with Kira's.

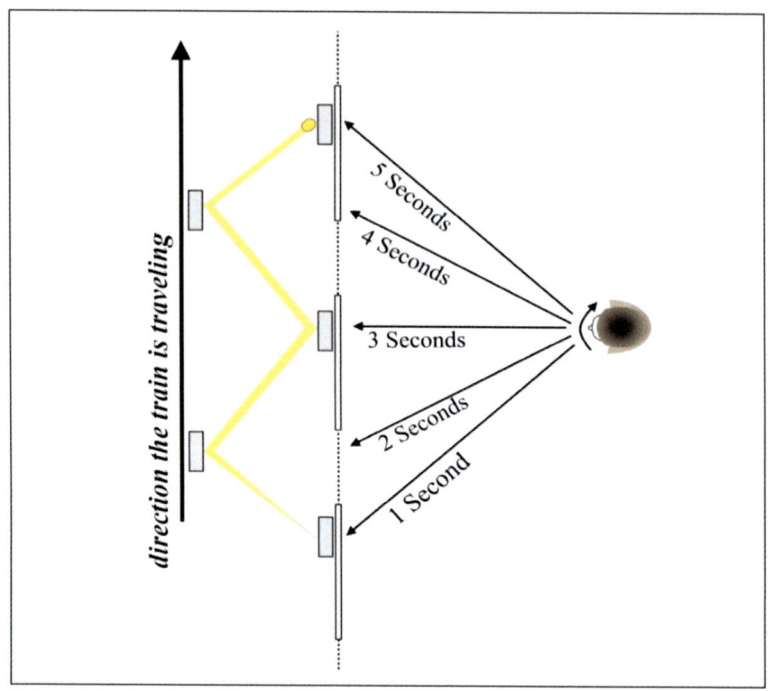

Figure 4-6b. While standing outside the train, Sora watches the photon of light bounce between the same two mirrors as Yori. However, since her reference frame is different from Yori's, the photon must travel a longer (zigzag) distance. Relative to Sora, the photon travels the cabin's width, plus the total distance the train travels forward during the 5-second interval.

Recall that Yori turns his head back and forth along with the photon as it bounces between the two mirrors. Also, recall that the distance the photon travels relative to Yori is short, merely the width of the train's cabin. Therefore, relative to Kira, who is also on the train, Yori turns his head *very quickly* from side-to-side. (Figure 4-6a) But how fast would Sora see Yori turning his head side-to-side?

Well, the longer zigzag trajectory that the photon travels, as seen by Sora, means that it is traversing a greater distance relative to her as well. Consequently, it takes the photon *longer* to reach and reflect on each mirror. Moreover, this increased distance also means that, relative to Sora, Yori is turning his head at a *slower*

130

rate than what Kira sees. Or, to put it another way, relative to Sora, Yori has to turn his head more slowly to follow the photon since she sees it following a longer trajectory. (Figure 4-6c)

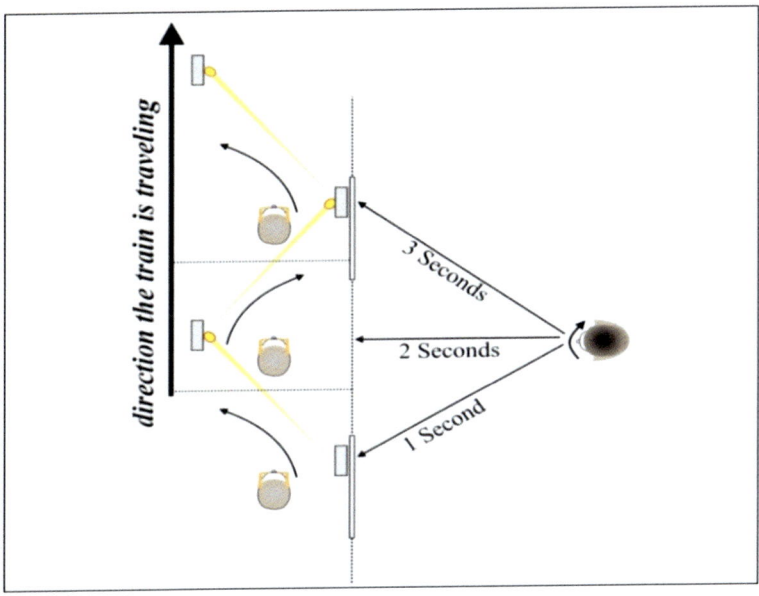

Figure 4-6c. The speed of light is constant. But the distance that Sora (outside the train) observes the photon traveling is longer; thus, it takes <u>longer</u> for the photon to travel that distance. Therefore, the speed at which she views Yori turning his head from side to side *must* be slower (i.e., takes <u>more time</u>) than what Kira observes.

Again, recall that speed equals distance divided by time, meaning that only two variables are required to define speed – *distance* and *time*. Therefore, since the speed of light is constant for all observers, if one of these variables increases, the other must decrease to compensate. So if from Sora's perspective at the station, the photon is traveling a *longer* distance, yet its speed remains unchanged, then time <u>must</u> slow down to compensate. As a result, as Yori turns his head to track the photon, he does so more slowly, relative to Sora. (Figure 4-6c) And indeed, the faster the train's velocity, the longer the distance the photon must travel; thus, the slower Yori's head will appear to turn (i.e., the slower time will flow onboard the train relative to Sora).

On the other hand, aboard the train, the distance the photon travels, relative to Kira, is just the 2.5-meter width of the train's cabin. Thus, the rate that time must flow to keep the speed of light constant is increased. Thus, Yori's head turns very quickly from side to side relative to Kira. (Figure 4-6a) The above is the only way these women, while in different frames of reference, can both measure the speed of light as 300 000 km/sec. Therefore, with the above explanation in mind, we can now summarize the full consequences of the Michelson-Morley experiment.

To start with: Maxwell's equations indicated that the speed of light was constant for all observers, _no matter their frame of reference_. And by seeing no diffraction pattern, the Michelson-Morley experiment confirmed Maxwell's results. Next, FitzGerald suggested that pressure from the aether was causing objects to contract or shorten in length. As a result, the light did travel slower, but the decrease in length caused by the aether's pressure made it appear to be traveling at the same speed. However, when Maxwell's equations are combined with FitzGerald's length contraction, Lorentz objected to the latter's conclusion. Speed equals _distance_ divided by _time_. Therefore, for the speed of light to be constant <u>to all observers</u> (_no matter their frame of reference_), length contraction cannot occur without time dilation. And so, for the first time in history, science was starting to see that space and time were not abstract concepts, background players on the "stage of life." Space and time were dynamic variables in the physical sciences. Still, it wasn't proving easy for the scientific community to let go of their belief in an aether.

For instance, Lorentz didn't call the slowing of time "time dilation" as we do today; instead, he called it _local_ time. He chose the term "local" because he didn't believe that time dilation was a real-world phenomenon. Lorentz's primary area of research was in atomic theory, which was where he thought the rules of his _local_ time applied (i.e., to the motion of the electron in a motionless aether). Consequently, Lorentz firmly held onto the belief of absolute time in the macro world, doing so based solely on an intuitive basis. In other words, if it is _now_ here, it must also be _now_ on the other side of the galaxy; in fact, it is _now_ everywhere in the universe. And yet, by refusing to accept that length contraction and time dilation occurred on the macro-scale, Lorentz ignored some unavoidable consequences.

A critical concept of relativity is that any two observers traveling at different uniform velocities will always obtain the same results for any identical experiment. A cup of coffee poured while flying in a plane will behave the same when poured at our kitchen counter. Indeed, in every activity we engage in, we count on the laws of physics to follow this principle. However, Lorentz's suggestion that "local time" only occurred on the atomic and not the macro level meant that experiments conducted at different velocities would yield different results. Such an idea presents enormous problems for all of science and technology.

For instance, what if the rules that govern chemical combustion, which operates on the atomic scale, only worked when your vehicle traveled at a certain speed? So instead of burning fuel at a predictable rate, the moment you reached 70 km/hour, several liters of fuel ignited instantly – every velocity with different laws of physics. If Lorentz's local time was correct, we could never know all the various laws of physics because there would be different laws depending on your speed. By accepting the idea that time varied on the atomic but not on the macro level, Lorentz rejected the notion of relativity.

However, to his credit, Lorentz was the first to connect length contraction to time dilation (i.e., the notion of space-time). This last concept turned out to be the crucial mental leap forward in explaining the "failure" of the Michelson-Morley experiment. As a result, although he incorrectly proposed the idea of "local time," his derived formulas equating length contraction and time dilation were correct. These formulas are still known as the Lorentz transformation equations in his honor. Moreover, Lorentz's absolute rejection of relativity turned out to be the catalyst that would change the nature of the debate. There were just ten years to go before Einstein would present his theory of special relativity, and the focus of the argument had switched. It went from the Michelson-Morley experiment and if the aether existed to the validity of relativity. Indeed, confirming the validity of relativity had been the underlying issue all along.

**Jules Henri Poincaré
(1854-1912)**

As FitzGerald and Lorentz debated the soundness of the Michelson-Morley experiment, the French mathematician, theoretical physicist, and philosopher Jules Henri Poincaré stepped in to voice his opinion. Poincaré took exception to Lorentz's rejection of length contraction and time dilation occurring on the macro-scale. Poincaré felt that for one to abandon the scientific principles of relativity represented a scientific relapse to pre-Galilean concepts. Therefore, in 1897, Poincaré published the paper *The Relativity of Space*. His goal was to prove that space and time were not absolute but obeyed the principle of relativity. And to accomplish this, Poincaré gave the simple example of two friends meeting at *Place du Panthéon* in Paris.

When two friends were done speaking one day, they agreed to meet in the exact location, *Place du Panthéon*, on the next day and at the same time. Then, exactly twenty-four hours later, the two friends meet at the very spot they were at the day before. But are they really in the same location as they were the day before? No. As Poincaré explained, since the Earth orbits the Sun, it has moved many thousands of kilometers away from the position it was in the day before. So, while the friends might agree that they're in the precise location they were in the day before, they certainly are not. Meaning that, although relative to them, they were in the same location, relative to someone observing from outside the solar system, they were not.

Poincaré's point was irrefutable. Lorentz's attempt to fit the results of the Michelson-Morley experiment into Newtonian physics on the micro-scale but not the macro was not going to work. Instead, there was a need to establish what Poincaré termed a *principle of relativity*. There needed to be a set of definite rules that could be applied in all frames of reference. Poincaré believed that this was the only way Galileo's notion of relativity could be respected. And with this final revelation, almost all the pieces of Einstein's theory of relativity were in place. In fact, as confirmation that Poincaré's suggestion had become necessary, scientists of his day had already started to notice problems with Newtonian mechanics. Problems that could not be ignored.

Le Verrier and Hertz

Before Isaac Newton, the field that had contributed most to the development of relativity was astronomy. However, during the 18th and 19th centuries, this changed as physics (most especially electrodynamics) had begun to figure more prominently. And yet, advancements in astronomy had not become irrelevant to the discussion. For example, something that astronomers began to notice when tracking the planet Mercury was that its orbit fluctuated, which came to be called a perihelion *shift*.* This slight (but persistent) discrepancy in Mercury's orbit had been calculated to be approximately 1 / 3,600th of a degree per century. (Figure 4-7) But, significantly, Newtonian mechanics could not explain why

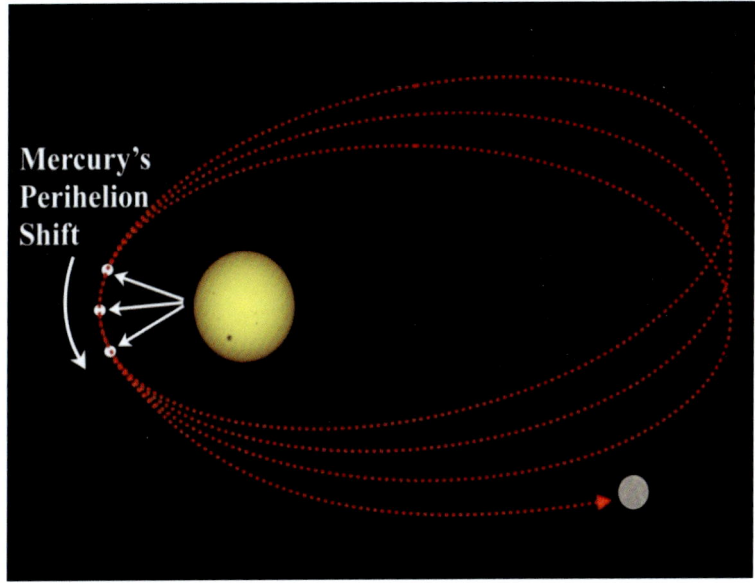

Figure 4-7. Mercury's perihelion, where its orbit brings it closest to the Sun, slowly changes with each orbit. This perihelion shift was not explicable using Newtonian mechanics.

---

* The word "perihelion" comes from the Latin where it meant "around the Sun." It is the closest point to the Sun in the orbit of a planet or comet.

it was occurring. However, Urbain Le Verrier thought he had found the answer. He suggested that perhaps the gravity of another planet, one even closer to the Sun, was perturbing Mercury's orbit. This same logic had been used before with great success.

Urbain Le Verrier was a French mathematician, astronomer, and director of the Paris Observatory. For many months in 1846, Le Verrier sought to explain why there was a slight deviation in the orbit of Uranus. In a spark of intuition, he realized that there must be another planet whose gravity was disturbing Uranus' orbit. Using math alone, he calculated the location of this new planet, then sent the coordinates to fellow astronomer Johann Gottfried Galle in Berlin for confirmation. And with that, the planet Neptune was dis-

Urbain Le Verrier (1811-1877)

covered. Now, using similar logic, perhaps there was an undiscovered planet inside the orbit of Mercury that was causing its perihelion to shift. Indeed, astronomers were so sure that this was the case that they even gave the unknown planet a name, Vulcan. Thus, astronomers of the era anxiously turned their telescopes heavenward to search for this unseen planet, hoping to be its discoverer. Yet, there was no planet Vulcan. The mystery of Mercury's perihelion shift was just more proof that something else was taking place, something beyond Newtonian mechanics. Then there was the extraordinary discovery of Heinrich Rudolf Hertz.

Heinrich Hertz (1857-1894)

In 1887, Heinrich Rudolf Hertz served as professor of theoretical physics at the University of Kiel, Germany. While conducting experiments in electrodynamics to verify certain theories of Maxwell, he was able to construct not just an electromagnetic transmitter but, even more importantly, a receiver. Yes, Hertz had heralded in the age of radio and television! However, though he took the first step toward our being able to binge-watch our favorite television shows, he had also noticed something bizarre. Hertz found that when electromagnetic *energy* impacted a metallic surface, it sometimes had the effect of causing a then-unknown subatomic particle to be emitted. Scientists initially called this particle a cathode ray; today, we know this particle to be the electron. The magnitude of Hertz's discovery was that he unknowingly detected what is now known as the photoelectric effect. When the correct frequency of light strikes an electron, the electron absorbs the energy, which allows it to break free from its atom. (Figure 4-8) But how this phenomenon occurred and was related to relativity would be a mystery until Einstein explained it in 1905.

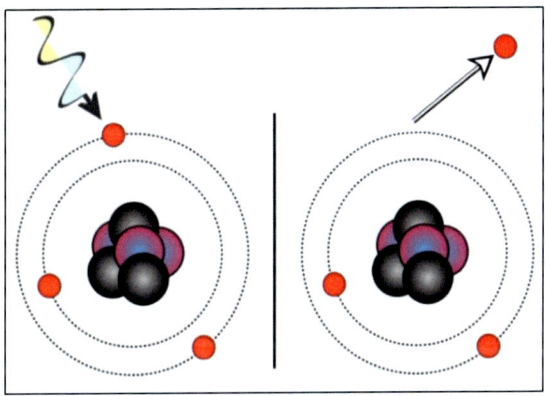

Figure 4-8. When the correct frequency of light strikes the orbiting electron of a metallic atom, the electron can absorb the light's energy and then use it to break free from the atom.

And with that, all parts of the Special Theory of relativity were now present. All that was required was for someone to assemble them into a coherent model and describe it mathematically. But before we explore how Einstein was able to do this, we must again acknowledge that the theory of relativity was not the sole conception of Einstein. As we've seen, many men and women of science, most of whom we have not mentioned, contributed pieces. Therefore, instead of viewing Einstein's theories as the handiwork of a single individual, it's best if we have the same opinion as Isaac Newton, who famously stated concerning his discoveries:

If I have seen a little further it is by standing on the shoulders of Giants.[*]

Sir Isaac Newton –

**Letter to Robert Hooke
February 5, 1676**

So, although more than 110 years later, we still marvel at what Einstein accomplished, it must be recognized that he wasn't alone in formulating these theories. Instead, Einstein's true genius was in his ability to assemble the many concepts of the theory of relativity into a working scientific principle. And now, we are also ready to do the same. We are now prepared to explore how a clerk who worked in the Swiss Patent Office could devise one of the most extraordinarily unconventional theories ever conceived.

---

[*] In fact, this phrase, most often attributed to Newton, was first spoken by 12th-century theologian and author John of Salisbury. He originally stated: "We are like dwarfs sitting on the shoulders of giants. We see more, and things that are more distant, than they did, not because our sight is superior or because we are taller than they, but because they raise us up, and by their great stature add to ours."

Chapter 5

On August 30, 1847, Hermann Einstein was born in Buchau, Germany; he was the 4th of six children. Although not much is written of his early years, when he was 14, Hermann attended a secondary school in Stuttgart. It's said he showed an aptitude for mathematics, though it's not believed he had any genuine interest in the subject. Unfortunately, because of family difficulties, the father of the 20th century's most renowned scientist would never be able to explore his academic talents. After earning a one-year certificate of completion

Hermann Einstein (1847-1902)

that exempted him from compulsory military service, Hermann was forced to drop out of school to help his financially struggling family. He took an apprenticeship as a merchant in Stuttgart. Later, upon completing his apprenticeship, he went into business with his cousin. And by all accounts, the two men were earning an extremely comfortable living selling mattress feathers.

Pauline Koch (1858-1920)

Pauline Koch was born in Stuttgart, Germany, on February 8, 1858; she was the youngest of four children. Pauline was from a moderately wealthy family of grain traders, was well-read, and had a talent for the arts; she excelled at playing the piano.[*] Pauline met Hermann when she was 18, and he was 29, and they married in 1876. Their first child Albert was born three years later, and two years after that, their last child, Maria. With Hermann's thriving mattress business and Pauline's well-to-do family, their marriage had all the markings of being a happy and affluent union. However, it wasn't long after the birth of Maria

---

[*] When Albert was six, his mother insisted that he learn to play the violin, something the young Albert at first despised. As an adult, however, playing the violin helped Einstein relax and think as he formulated his theories.

that Hermann decided to expand his business, a fateful decision that would end up causing his young family years of economic hardship.

During the late-19th century, cities throughout North America and Europe were replacing oil streetlamps with electrical ones. Hermann Einstein's youngest brother, Jakob, having just completed school as an electrical engineer, was eager to start up his own utility company. However, Jakob lacked the funds to do so on his own. So he convinced his brother Hermann to abandon his successful mattress business in Ulm and to back him financially. The brothers would be installing the wiring required to light the streets of Munich. Jakob would handle the engineering and technical aspects, while Hermann would act as business manager and investor. In 1882, when Albert was just three years old, Hermann took the chance and moved his family to Munich.

With lucrative city contracts enticing many into creating start-ups, the utility business was highly competitive. As a result, despite their most sincere efforts, the Einstein brothers lost their bid to light Munich. Still, this early setback didn't dampen their enthusiasm. In 1894, they relocated their company and families to Italy, hoping to succeed in that emerging market. And yet, the company continued to lose money. The brothers refused to abandon their venture and struggled to make their business profitable for the next decade. However, in the end, after amassing a large debt, they were forced to liquidate their assets to pay creditors, which proved to be the final obstacle that they could endure together.

In 1896, the Einstein brothers finally separated, with Jakob Einstein taking work as an engineer with an established firm. But Hermann Einstein still refused to give up. Against the wishes of a now 17-year-old Albert, Hermann continued to try and establish his utility company. Even after the second company folded, Hermann Einstein persisted. He formed a third utility company, but one that focused on constructing power stations, and this time his gamble appeared to pay off. Hermann had finally begun to meet with success, though it came at a high cost.

For roughly two decades, Hermann Einstein had to deal with constant financial worries and business failures, and these stresses had finally taken a toll on his health. Soon after starting his third company, Hermann began experiencing issues with his heart. Sadly, in October of 1902, Hermann Einstein died at 55. Had he lived just three years more, he would have witnessed what is known in science as the "miracle year," the year when Einstein published his four revolutionary scientific papers.[*] But, like Isaac Newton's father, Hermann Einstein never knew the international recognition that his son eventually acquired.

After her husband's death, Pauline decided to move in with her oldest sister, Fanny. The latter had married Rudolf Einstein, a first cousin of Hermann on his father's side.[†] Pauline stayed with her sister in Hechingen, Germany, until the war in 1914. After the war began, she moved in with her brother, Jacob Koch, and his family in Zürich. By this time, however, Pauline had now become incredibly ill. She was diagnosed with abdominal cancer. For several years she fought to overcome the disease until, in 1918, she had to seek long-term care at a sanatorium. Still, although she received treatment for a year, Pauline's health only declined. Finally, in 1919, Albert brought his mother into his home in Berlin. Pauline Einstein died on February 20, 1920, surrounded by relatives.

Albert and his younger sister, Maria, were very close. (Figure 5-1) As children, she was Einstein's best, if not only, friend. Maria was also very scholastically proficient, although it doesn't appear she shared her brother's interest in science. In 1909, Maria was awarded her Ph.D. in romance languages and literature. A year after graduating, she married Paul Winteler. He was the headmaster's son of the school that Albert attended prior to taking his college entrance exam. By 1922, the same year Benito Mussolini came to

---

[*] The first paper, "On a Heuristic Viewpoint Concerning the Production and Transformation of Light" (for which he won the Nobel Prize), described the photoelectric effect. The second paper was on Brownian motion. It described the motion of particles in gases and liquids, as first observed by Robert Brown in 1827. The third paper was on special relativity and explained why light had a constant speed in all frames of reference. Finally, in November, he published the paper "Does the Inertia of a Body Depend upon Its Energy Content?" It was in this paper that, for the first time, the relationship between matter and energy was shown through the celebrated equation $E = mc^2$.

[†] Fanny and Rudolf Einstein were also the parents of Einstein's cousin and second wife, Elsa Einstein.

Figure 5-1. Albert, age 14, with his 12-year-old sister, Maria. (1893)

power, Maria and Paul lived in Italy. And they remained there for 17 years until the tide of war began to rise again in Europe.

The seeds for the Second World War were sown after the first, in 1919, with the Versailles Treaty. But indications that a second war was imminent were not seen until 1927 when Mussolini dismantled the Italian democracy and installed himself as a dictator. By this time, Einstein had been a professor at the Berlin Academy of Sciences for some 14 years. Then in 1930, he received state permission to lecture for a few months each winter in the United States. After completing one such trip in 1933, Einstein learned that the German Nazi party had finally come to power. Before his ship landed in Antwerp, he had penned his resignation letter to the academy. Einstein then promptly returned to the United States. Finally, in 1937, while on a state visit to Germany, Mussolini announced that the two countries were now allies, even boasting that Europe would revolve around the "axis" of the German-Italian alliance. However, despite this ominous pact, Mussolini was reluctant to enact race laws in Italy as had already been done in Germany. But this changed in July of 1938 with the publication (in Italian) of *The Manifesto of Race*.

Among other things, *The Manifesto of Race* stated that Italians were direct descendants of the Aryan race and that all who were not were of inferior heritage. Using the manifesto as a basis, by September of 1938, Italy passed its first anti-Semitic laws. These laws stripped Jews of their Italian citizenship and removed them from positions within the government, academia, and all other professional careers. Albert urged his sister and her husband to join him in the United States. But while both applied in 1939, only Maria was accepted. Paul was not allowed to immigrate, presumably, due to some unknown health problem. And while Paul was safe from the rise of anti-Semitic rhetoric, since he was not Jewish, Maria was not. As a result, Maria was forced to leave Italy to live with her brother and his second wife in Princeton, New Jersey. Paul returned to Geneva, Switzerland, to live with relatives. It was only after the war, in 1946, that Maria and Paul began making plans for her to return to Europe. However, in that same year, Maria suffered a severe stroke and was thereafter bedridden. She eventually died in 1951; her husband Paul died in Switzerland a year later.

Albert Einstein
(from 1879 to 1905)

On March 14, 1879, Albert Einstein was born in Ulm, Germany. And although he would become the most celebrated scientific mind of the 20th century, this was not apparent to those who knew him as a child. At birth, for example, there was immediate concern with his physique. His mother was alarmed by how "lopsided" his head was, whereas his grandmother exclaimed that he was "too fat." Yet, while these concerns were more maternal vanity than a valid ailment, a more legitimate fear arose three years later.

Around 6-9 months, an infant will start to babble random sounds; by 14 months, the child will begin to speak words. But Einstein didn't follow this normal progression, which concerned his parents, who feared that their son might be developmentally challenged. Einstein did not speak until he was nearly three years of age, though, in truth, this is not unheard of for some children. Still, Einstein did struggle when it came to language. He didn't master his native tongue of German until he was nine years of age, about two full years behind the average seen in most children. Happily, however, his parents' fears proved unfounded. In time, their son learned to speak two languages: German and English.

Another recurring opinion of young Einstein (by those involved with his education) was that he was not an impressive student. Even in college, he didn't seem to impress any of his instructors. Although his grades were always good to excellent, most of his teachers classified him as lazy and even insubordinate. This latter quirk undoubtedly being the primary cause of their unflattering observations. Many modern educators theorize that Einstein's insubordinate attitude may have been because he was bored with how and what he was taught. Their assumption comes from an experience in Einstein's early childhood.

When Einstein was 11, his family began hosting a poor college student once a week for a meal; a man named Max Talmud, a 21-year-old medical student. Talmud would bring his science textbooks and discuss them with the utterly fascinated Einstein during his regular visits. Talmud visited the Einstein family for ten years. In this environment and with this topic, young Albert flourished; it's believed that this was when he learned to love science. Indeed, when he was only 16 years old, Einstein wrote his first scientific paper, *On the Investigation of the State of the Ether in a Magnetic Field.* This paper demonstrated that Einstein did indeed have an interest in (and an ability to)

---

* It is of interest to note the title of Einstein's first scientific paper. At just 16 years of age, Einstein had already begun to ponder the question that would lead him to his special theory of relativity. That question was: What would it be like to chase, and then catch up to, a beam of light on a bicycle? This unassuming question of a teenage boy inspired one of the most revolutionary scientific theories in history.

assimilate advanced topics. It also showed where his scientific interests were leading him. Thus, it may well have been that Einstein was simply bored throughout his school years.

In 1894, when Hermann moved his business to Italy, Einstein remained with relatives in Germany to complete high school. But, unhappy with being left behind, young Albert dropped out of school, though he had a plan. He believed he could challenge the entrance exam and begin college in Zürich, Switzerland, just a few hours away from his family's new home in Pavia, Italy. Unfortunately, he failed the exam on his first attempt in 1895. Still, he persisted, a trait he undoubtedly acquired from his father. But more importantly, the university's principal was impressed with Einstein's mathematical aptitude. Therefore, the principal arranged for Einstein to spend the next year studying to retake the exam at one of the local high schools. At the same time, Einstein was allowed to live with the school's headmaster, Professor Jost Winteler. Einstein completed high school while preparing to retake the university entrance exam. Finally, in October of 1896, Einstein passed the entrance exam to the respected Swiss Federal Institute of Technology on his second attempt. (Figure 5-2)[*] Nonetheless, upon starting college, Einstein was confronted with the same problem he had long endured for much of his academic career.

Figure 5-2. The Swiss Federal Institute of Technology in Zürich, Switzerland – the college that Albert Einstein attended.

[*] Image credit: http://commons.wikimedia.org/wiki/User:Roland_zh

Although he was now learning at the level that he thought he would enjoy, Einstein continued to find school uninteresting. And, also unfortunate, he didn't refrain from telling his instructors how he felt. What's more, Einstein regularly skipped classes, preferring to remain at home to read the latest scientific papers. Still, to the frustration of his instructors, Einstein never failed an exam. After four years of attendance at the Swiss Federal Institute of Technology, Einstein graduated with a degree in physics. However, much to his surprise, he quickly realized that merely having the diploma was no guarantee of finding work. The many years he had spent criticizing and being condescending toward his instructors came with a terrible price. No one would give him a letter of recommendation; without which, it would be almost impossible for Einstein to get a teaching job. So, for almost two years, Einstein was forced to subsist solely on temp work. Then a friend got him a steady job as a patent clerk at the Swiss Patent Office in Bern.

Mileva Maric (1875-1948)

One year after starting work at the patent office, Einstein married a former classmate named Mileva Maric. Eventually, they would have three children: Lieserl, Hans Albert, and Eduard.* Today, there is much debate about how much Einstein's first wife assisted him in formulating his theory of special relativity. Although Mileva was unable to pass her exit exam to receive her 4-year degree, her college transcript shows that she did complete courses in differential calculus, mechanics, and theoretical physics. And

---

* Lieserl Einstein was born out of wedlock in January 1902. It's believed she lived with Mileva's parents but died of scarlet fever at a very young age; however, nothing of certainty is known as to the fate of Einstein's oldest daughter. Hans Albert followed in his father's footsteps. He also studied at the Swiss Federal Institute of Technology, earning his Doctorate in Technical Sciences. Hans Albert had a distinguished career as a civil engineer, and even today, his work is often referenced. Einstein's youngest son, Eduard, developed schizophrenia when around 20 years old and spent much of his life in a psychiatric clinic in Zürich.

from correspondence exchanged between her and Einstein, some suggest she may have assisted him in mathematically deriving his special theory of relativity. This opinion, however, cannot be confirmed.* And there is a second related question regarding Einstein at this busy point in his life that also remains unanswered. For the seven years he worked at the patent office, attended graduate school, and cared for a family, how could he develop such a profound theory in his spare time?

Many explanations are given about how Einstein found the time to formulate his theory while juggling such a busy schedule. And perhaps they all have a grain of truth. But, in all honesty, many overlook one thing that proved invaluable to Einstein in formulating his theory: most of the pieces that made up special relativity were already in place before 1905. Now, this last fact in no way diminishes Einstein's intelligence – indeed, all of *general* relativity was Einstein's conception. Yet, in truth, Einstein's intellect was not enormously beyond that of many other scientists of his era; some of the most outstanding scientists of the 20th century were alive at the same time.† Thus, another physicist would have certainly (eventually) assembled special relativity had it not been Einstein. However, intellect aside, Einstein also possessed another personality trait that also proved to be an asset: his insubordinate and determined nature.

It was never a concern to Einstein what others thought of him, nor did he hesitate to challenge authority. Indeed, this "insubordinate" nature appears to have been a dominant personality trait found in several of the scientists considered. Aristotle challenged Plato's method of reasoning, Galileo

---

* The debate around how much Mileva assisted Einstein in formulating his special theory of relativity centers primarily on comments made between them in love letters. But these comments, without knowing their exact context, can be interpreted rather subjectively. Other details often cited as proof can be similarly disputed. So, while she may well have aided Einstein, there's no conclusive proof either way.

† At his death, Einstein's brain was removed by Dr. Thomas Harvey, who performed the autopsy. It was eventually divided into 240 sections. These sections eventually found their way into the hands of several neuroscientists. Despite possessing several distinct attributes, it was determined that Einstein's brain was average.

challenged church dogma, and Émilie du Châtelet challenged cultural norms. Despite the consequences, they all shared the insubordinate disposition of not hesitating to stick to what they believed when they thought they were correct. Einstein had this trait as well. Certain aspects of Einstein's theories, such as light behaving as a particle and the absolute rejection of the aether, directly challenged the beliefs of such distinguished scientists as James Clerk Maxwell. While other scientists of Einstein's day who examined the same scientific evidence that he did, refused to consider what that evidence was implying. Perhaps they refused to do so because they were afraid to challenge the accepted scientific dogma. But Albert Einstein was not afraid to do so, and his gamble, although it could have disgraced him, paid off. And with that, we are ready to return to 1905 and the single question that Einstein had pondered for more than ten years.

The Thought Experiment

So, what would it be like to chase and then catch up to a beam of light? This question preoccupied Einstein for more than a decade. To understand how the answer led him to special relativity, we can conduct the same *thought experiment*. Start by imagining a bright beam of light streaking through the air a meter above the ground. Besides the beam is a young Einstein riding a bike. Of course, the light beam moves at the *speed of light*, represented by the lower-case letter **c**. Finally, there is young Einstein. He is peddling furiously, has a strong wind at his back, and, at some point in his ride, he reaches the speed of light. Now here is the all-important question: As he peers over at the beam of light traveling beside him, <u>what does he see</u>?

According to Maxwell's equations, light is a continuously alternating electromagnetic field. (Figure 4-2a) The advancing electric field produces a magnetic field and vice versa. Moreover, this oscillation occurs at a constant speed, **c**. But as we learned From Galileo's ship experiment, when two objects share the same velocity, they are also motionless relative to one another. Therefore, once Einstein has caught up to the beam of light, one possibility is that he will see an unchanging (non-oscillating) waveform. We can visualize this by imagining a surfer paddling out from shore.

A surfer paddles out into the ocean. As he does so, it is clear that the waves are all traveling inward toward the beach; this occurs because his outbound frame of reference opposes the shore-bound waves. However, once he catches a wave and rides it toward the shore, the waves will appear to be standing still. Because he now moves at the same velocity and in the same direction as the wave he is riding, all the waves around him appear motionless. Consequently, this was Einstein's first conclusion. If he could travel at the speed of light and saw a beam of light traveling beside him, it would appear as a static, unchanging waveform. (Figure 5-3) However, there are problems with this conclusion.

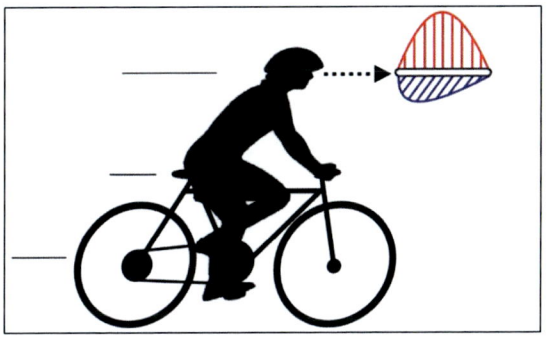

Figure 5-3. In Einstein's thought experiment, as the cyclist reaches the speed **c**, the beam of light that he is traveling alongside appears to be a static (non-oscillating), stationary (motionless), electromagnetic waveform.

Earlier, we discussed how Oersted saw the compass' needle immediately deflected when he turned on the current beside it. Then, just as quickly, the compass returned to normal. So, switching the current on and then off causes the compass to swing back and forth. From this, we know that only a *continually* oscillating electric/magnetic field will produce an electromagnetic wave. Therefore, Oersted's experiment agreed with Maxwell's equations. When an electromagnetic wave stops oscillating, it no longer exists. So, Einstein's first conclusion (that once he reached the speed of light, the light wave would be standing still) violated Oersted's experiment and Maxwell's equations.

Another possibility was that the beam did become motionless upon his reaching the speed of light and, therefore, would no longer exist <u>relative to him</u>. So, once young Einstein reached the speed of light, he would see nothing. However, the problem with this conclusion is that the beam of light would still exist to anyone standing relatively motionless. It is a contradiction for the same beam of light to exist to a stationary person but be nonexistent to someone in motion. This conclusion would violate the principle of relativity.

These conflicting results (i.e., a light wave that, though not oscillating, continued to exist and could be seen, as opposed to a beam of light that does/does not exist in differing frames of reference) were a problem. But, after ten years of pondering the issue, Einstein believed he had finally found the answer. And that answer resided in two conclusions; these were:

1) For light to exist, it can never be motionless. In effect, light *must* always be oscillating and traveling forward at a constant speed of 300 000 km/s; and

2) No object possessing mass can reach the speed of light. The reason is that if it could reach the speed of light, then, relative to that object (or person), the light wave would be standing still, which would violate the first conclusion.

And with that, having reached the two conclusions above, Einstein was now ready to formulate his *special* theory of relativity.

Chapter 6:
The Special Theory of Relativity

Space-Time

Einstein believed that he had finally resolved what could and could not happen if he were to travel at the speed of light. He was now ready to assemble the many scientific fragments that would come to make up his theory of special relativity. He began by devising two postulates, which formed the core of his argument. These postulates were based on Galilean relativity and Maxwell's equations, respectively.

Postulate 1 - The principle of relativity *always* holds. Within any inertial frame of reference, no experiment can be conducted that could reveal if one was moving at a uniform velocity or was at rest.

Postulate 2 - In *every* inertial frame of reference, the speed of light is constant and independent of the motion of the source that emits it.

And just like Aristotle's theories of motion, these postulates of Einstein also implied definite consequences. The first and most apparent was the elimination of the aether.

Einstein did not attempt to fit the results of the Michelson-Morley experiment into the old-world view of a universal, omnipresent aether. Instead, he took the results of that experiment as evidence that the aether didn't exist. Then, using an explanation similar to that found in Figures 4-6a and 4-6b, Einstein showed that time dilation was a real-world, macro-scale phenomenon. Thus, the final result of these conclusions was that the speed of light remained constant in all inertial frames of reference (i.e., at all velocities). But to fully appreciate what this implies, consider the formula for speed.

distance ÷ time = speed

151

The above formula tells us that an object's speed is based on two essential factors: the distance the object has traveled and how long it took the object to travel that distance. Without these two figures, the speed of an object cannot be determined. So, let's insert the speed of light into the above formula.

300 000 kilometers ÷ 1 second = 300 000 km/s

Now, imagine a rocket traveling at 10% the speed of light: 30 000 km/s. While at this speed, the ship's passenger turns on his forward lights. According to Newtonian mechanics, the two velocities (i.e., that of the rocket and the beam of light) should combine using addition. It would be as if you were traveling at 55 km/hr and threw a rock forward from the car window at 15 km/hr. The rock's total velocity would be 70 km/hr. Therefore, *according to Newtonian mechanics*, a rocket traveling at 30 000 km/s that emits a light beam should result in the beam of light traveling 330 000 km/s. But this is faster than the speed of light and, thus, is not allowed. So what's the mathematical answer?

Maxwell's Equations are correct; the speed of light is constant in every frame of reference. That's why the simple addition of velocities can not apply to light; light's speed is always constant, regardless of the motion of the object emitting it. However, there is a second variable within the formula for speed: *Time*. To keep the speed of light constant, the rate at which time flows *must* vary. And this was Einstein's greatest insight! Hence, in our example above, time will slow *relative to the rocket's frame of reference*. Indeed, relative to anyone aboard the rocket, time is moving slower. Hence, given that the rocket's velocity is 10% of the speed of light, the rate that time flows for those on the rocket must have slowed by a corresponding amount. Inserting these numbers into the formula for speed gives us:

330 000 km ÷ 1.1 seconds = 300 000 km/s

As the formula shows, by slowing the rate of time, the net effect is that the speed of light remains constant.* And this is the core concept of special relativity.

- *The constancy of the speed of light in all frames of reference makes time dilation a necessary aspect of reality.*

And yet, Einstein also realized that time dilation was only half of the solution in maintaining the constancy of the speed of light; the other half was length contraction.

Consider two observers. The first is our passenger aboard the rocket, and the second is standing motionless relative to the spacecraft. (Figure 6-1) Both also have three green measuring rods that, for the sake of argument, are all 100 000 km long. All three rods placed end-to-end equal the distance light travels in one second. Next, the person in the rocket blasts off and reaches a uniform velocity of 10% of the speed of light. He then turns on his forward lights. Finally, precisely 1 second later, both observers separately and *instantaneously* use their individual measuring rods to determine how far the light has traveled. Remember, for both individuals, the speed of light is constant. Thus, the light has traversed the same distance in both inertial frames. Let's begin with Observer #2.

Observer #2 stands motionless compared to the rocket. As he sees the forward lights turn on, the rocket is at position **A**. One second later, the beam of light reaches position **C**. Using his green measuring rods, Observer #2 measures from point **A** to **C**. As expected, the distance the light has traveled is 300 000 kilometers; all three rods laid end-to-end. (Figure 6-1) Now examine the situation from Observer #1's perspective onboard the rocket.

Observer #1 agrees with Observer #2; the forward lights came on at position **A**. Observer #1 also concurs that, one second later, the beam of light has reached position **C**. However, unlike Observer #2, Observer #1 is in relative motion; the rocket's velocity is 30 000 km/s. As a result, after one second, the rocket

---

* The above formula for time dilation is overly simplified but illustrates the point.

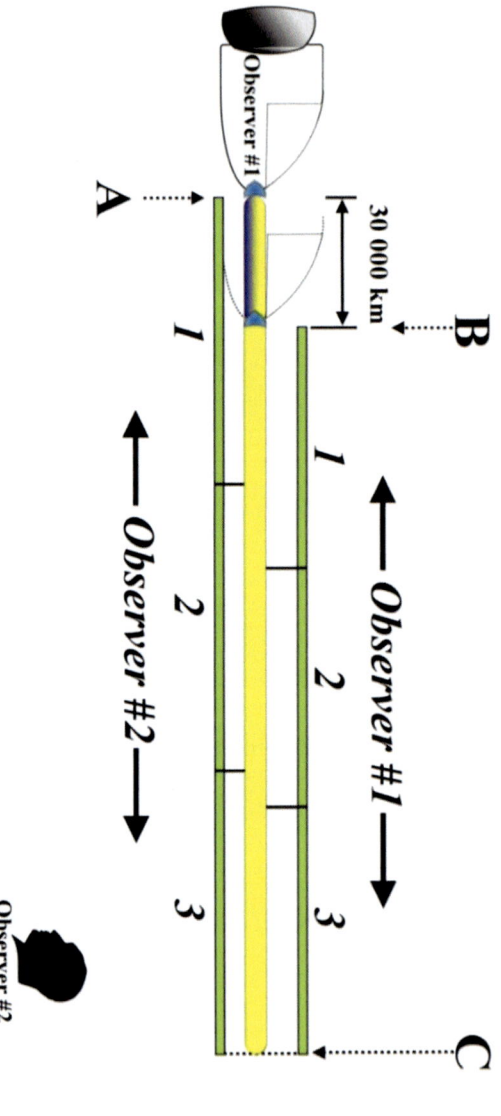

Figure 6-1. Relativity states that there is no difference between an observer in uniform motion or one at rest; the laws of physics remain the same for both observers. For both Observer #1 (onboard the rocket) and Observer #2 (standing motionless outside), the light beam reaches the same point. Observer #1 (using his three green measuring rods) measures the total distance the light has traveled in 1 second from point **A** to point **C**. However, Observer #1 is no longer at point **A** after 1 second. Therefore, he (using his three green measuring rods) measures the distance the light has traveled from point **B** to point **C**. Length contraction in the direction of travel has caused the rocket, Observer #1, and his measuring rods to contract; thus, the speed of light remains constant.

154

(along with Observer #1) is no longer at position **A**. Therefore, when Observer #1 measures with his rods how far the light has traveled, he measures from position **B** to **C**; all three of his rods laid end-to-end. (Figure 6-1) But for Observer #1 to measure an identical distance traveled, length contraction must have occurred; the rods that Observer #1 uses have gotten shorter. And yet, this outcome suggested an even more profound consequence, a consequence that no one prior to Einstein had ever considered. It appeared that space and time were somehow interconnected.

Once Einstein accepted that space and time *cooperated* to keep the speed of light constant, his next conclusion was obvious. Space and time are interrelated. Like a fabric that was intricately woven together, space and time were a single cloth that adorned matter and energy. Thus, before Einstein, the universe (space) was believed to have only consisted of three dimensions: length, width, and height. But with special relativity, space now was combined with time. So we now speak of the universe being made up of four dimensions: length, width, height, and *time*. And just as electricity and magnetism became electromagnetism, now space and time became *space-time* or the *space-time continuum*.* We can better illustrate this by imagining a friend asking you to pick him up at the airport.

A close friend is returning from a trip to the Serengeti and asks you to pick him up at the airport, so he gives you the airport's address. The airport's address corresponds to its measured position in space. Yet, instinctively, you automatically ask another question: What time does your plane arrive? In order to locate any object within the space-time continuum, including your friend at the airport, one must know its physical location *and* at what time to look for it. If you go to the airport to pick up your friend at 4:00 pm, but his plane does not arrive until next Tuesday – although you will be at the correct spatial coordinates – your friend will not

---

* A *continuum* is defined as the continuous and systematic interrelationship of two or more things. And while each is individually distinguishable, they don't exist apart from one another. Therefore, the space-time continuum is the continuous interrelationship of 3-dimensional space with the single dimension of time.

be there.* Space and time are linked. And still, time dilation, length contraction, and a single space-time continuum were not the only oddities of the special theory of relativity. Einstein suggested another effect that must also be occurring, one which he called *Relativity of simultaneity*.

Imagine two lightning strikes occurring a great distance apart; in fact, the two locations are three light-minutes apart from one another. (Figure 6-2) Next, imagine standing between these locations at position **B**. At your position, it takes 2 minutes for you

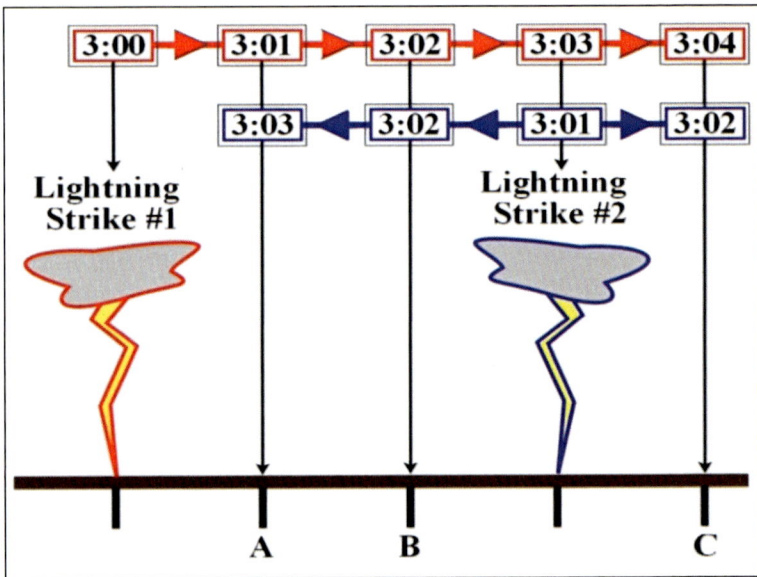

Figure 6-2. Depending on your location relative to the lightning strikes, the time that you perceive them as occurring changes. This effect is known as *Relativity of Simultaneity*. For example, persons located at points **B** and **C** will perceive lightning strike #2 as occurring simultaneously, 3:02; however, they will not perceive lightning strike #1 as occurring at the same time as lighting strike #2 because their spatial distances from lightning strike #1 are different.

---

* It's not unlike someone telling you to meet them at the Burj Khalifa (the tallest building in the world) at 5 pm. Although you may have the longitude, latitude (corresponding to length and width), and what time, it would be extremely difficult to locate them if you don't know what floor (height) they are on. The fabric of the universe is defined by all four: length, height, width, and time.

to see the flash of light originating from lightning strike #1. So, even though lightning strike #1 occurs at 3:00, by the time you see its flash at position **B**, it's 3:02. However, it takes only one minute for light from lightning strike #2 to reach position **B**. So while lightning strike #2 occurs at 3:01, by the time you see its flash at position **B**, it is also 3:02. As a result, even though the two lightning strikes do not occur simultaneously, because of your location at position **B**, it appears to you as if they do. Next, imagine persons at each of the three positions shown in Figure 6-2 (positions **A**, **B**, and **C**). When they finally came together and compared the times that the two lightning strikes occurred, they would all have different answers.

Relativity of simultaneity occurs when separate observers have different frames of reference – either spatial (their position) or inertial (their motion). As a result, neither can say the exact timing of the same event with absolute certainty. In other words, although we'd like to believe that lightning strike #1 occurred at 3:00, the *thing* that tells us that a lightning strike has occurred (i.e., the flash of light) is as valid at position **A** as it is at **B**, as it is at the exact location of the strike. As a result, special relativity forever ended the idea of action-at-a-distance (i.e., instantaneous effects).

Now, recall from chapter 3 that Newton believed that the force of gravity was both universal and instantaneous. Consequently, if the Sun were to disappear right now, the Earth, no longer being held in place by its gravity, would *immediately* fly off into space. Yet relativity of simultaneity changed that. Rather than being universal and instantaneous, the *effect* of gravity now had a maximum speed limit – the speed of light. And the latter idea challenged thousands of years of human thought. Recall that Lorentz rejected relativity of simultaneity on the intuitively based premise that *now* is *now* everywhere. But let's consider why Lorentz and the idea of action-at-a-distance are wrong. First, we know that the speed of light is fixed, now think of what would happen if the Sun disappeared in the following scenario.

Take a moment to look at a nearby clock. At the exact time that you just saw, the Sun vanished. If this had indeed happened, historians would chronicle the end of the world as occurring at the time you just saw on the clock. But would they be correct? No. The

Sun is 150 million km away. And since light (and gravity) is not instantaneous, it would take $8^1/_2$ minutes before we would have *any* indication that something was wrong. Therefore, the end of the world (i.e., the time that the Sun actually disappeared) would have occurred $8^1/_2$ minutes before the time you saw on the clock. But let's further imagine that, at the time you saw the Sun vanish, you sent a distress call via a radio signal to your good friend living on Mars. Coincidently, he's looking at the Earth and the Sun through his telescope as you send the radio transmission. The question is: What does he see and hear, and when does he do so?

It takes light from the Sun a little more than 12 minutes to reach Mars. Therefore, $8^1/_2$ minutes after the Sun has vanished, as the Earth begins to fly off into the darkness of space, it is still a sunny day on the Martian equator for another 4 minutes. Relativity of simultaneity means that no information can travel faster than the speed of light.* Therefore, 4 minutes later, the same 4 minutes it takes your radio signal to reach Mars, your close friend now watches as the Sun disappears from the sky. But here is what's strange: Your friend on Mars will hear your distress call on his radio at the precise instant the Sun disappears. So, to him, the disappearance of the Sun *and* your distress call happened simultaneously. Though, in reality, these events occurred $8^1/_2$ minutes apart.

It takes light from the Sun 43 minutes to reach Jupiter. So, the Sun's disappearance would have occurred 43 minutes ago, affected the Earth 34 minutes ago, and affected Mars 30 minutes ago. But the phenomena telling you that these events have occurred all propagate at the speed of light. Therefore, the Sun's disappearance, the distress call from the Earth, and a distress call from Mars will all appear to happen at the same moment to someone on Jupiter. Lorentz's belief that *"now"* is *"now"* everywhere can't possibly be true. As Einstein correctly deduced, all physical phenomena require *time* to propagate, even light and gravity. And the speed they both propagate at is the speed of light. They are not instantaneous.

---

* Information means any indication (i.e., light, gravity, protons, etc.) that could indicate that the Sun had disappeared.

Consequently, based solely on when we perceive a phenomenon, *"now"* varies depending on location.* In other words, what we perceive as *"now"* (the present) is the arrival of information to our specific location at this very instant. For someone standing a kilometer away, the sound of the barking dog hasn't reached them yet, nor have they felt that cool breeze that you have just experienced. Their "now" is not the same as yours.

So, Einstein's special theory of relativity had successfully described how time dilation and length contraction worked. And it also replaced the idea of action-at-a-distance, with relativity of simultaneity. These concepts forever changed how the universe was viewed because space and time could no longer be treated as individual and separate. As Einstein himself stated, the three dimensions of space were inexorably combined with time; the space-time continuum. However, Einstein was not finished. While Lorentz and FitzGerald were the ones who initially proposed time dilation and length contraction, the next aspect of special relativity was conceived of entirely by Einstein.

Mass-Energy Equivalence

Until the 19th century, Newton's laws of motion could describe the movement of all the then-known objects in the then-known universe. But with Maxwell's Equations proving the constancy of the speed of light and the Michelson-Morley experiment disproving the aether, Newton's laws were no longer applicable to extremely high-velocity objects. To understand why, recall the debate between Newton and Hooke regarding light, whether it was a particle or wave.

---

* All physical phenomena require time to propagate. So let's say you're using the light to tell you when an event occurred. Since light has a definite speed, your spatial position relative to the event will affect the time you perceive the event as occurring. But let's say you are using the lightning's electric shock to tell you when the event occurred. (Figure 6-2) A person standing 1.5 meters away from the lightning strike would be electrocuted earlier than a person standing 15 meters away. Again, all phenomena require time to propagate. And you cannot tell if an event has occurred until evidence of its occurrence (e.g., the light, the electricity, or some other physical phenomena) has reached you. Depending on where you are located relative to the event, the evidence of its occurrence will reach you at different times.

Let's assume Newton was correct, that light existed as many distinct particles. And we know that Maxwell's equations proved that the speed of light was constant – 300 000 km/s. When combined, these two statements would mean that it would be fatal when a particle of light struck you. Because according to Newtonian mechanics, as the velocity of an object increases, so would its momentum ($p = m$v). Therefore, if light were composed of *massed* particles traveling at 300 000 km *per second*, they would strike with such force that all life on Earth would already have gone extinct.[*] Therefore, Newton's idea that light was a particle had to be incorrect. Yet, there was a similar problem with Hooke's theory of light.

Hooke was correct in theorizing that light propagated as a wave; however, he also believed that the aether transmitted light. But if the aether transmitted light as a pressure wave, Newton's 2nd Law of motion would still apply, except it would apply to the aether itself. Meaning that the aether particles would be vibrating all around us *at the speed of light*, creating the same issue as a fast-moving light particle. So, what was the answer? Was light a wave or a particle?

Newton's laws of motion worked exceptionally well for everyday velocities. For this reason, it's still taught in engineering disciplines at all levels. But Einstein realized that Newton's laws of motion were inadequate when considering objects traveling at extremely high velocities. As the speed of an object increases to the point where it becomes necessary to measure its rate in kilometers *per second*, Newton's laws of motion break down. Therefore, Einstein's solution was to split the difference between

---

[*] This is the greatest danger faced by any manned mission to Mars. In addition to radiation, the Sun ejects massive amounts of high-energy particles (primarily protons and electrons), moving at speeds of up to 90% the speed of light. At this velocity, outside of Earth's protective magnetosphere, these *massed* particles can ravage living cells. High-speed subatomic particles can cause memory loss, damage eyesight, and break DNA chains, causing unwanted mutations, premature aging, and cancer. In fact, during the Apollo moon missions, astronauts reported seeing bright flashes of light when their eyes were closed. High-speed subatomic particles colliding with either their retinas or directly with their brains were interpreted as bright flashes of light. These flashes of light were so intense that they were even disrupting the astronauts' ability to sleep.

Newton and Hooke. Instead of declaring that light was *either* a particle or a wave, Einstein said it was both.

In March 1905, in his paper *On a Heuristic Viewpoint Concerning the Production and Transformation of Light*, Einstein described light as a **massless particle** that carried the electromagnetic **wave**. This reimagining of electromagnetism is the reason we now conceive of light as having a dual nature. At times, light behaves as a particle, which we call the photon.[*] While at other times, it behaves as a wave. As a result of redefining light as a *massless particle*, it was no longer obliged to obey Newton's laws of motion, which only applied to objects possessing mass. And with his rejection of the aether's existence, while maintaining light's wave-like nature, Einstein ensured that light still obeyed Maxwell's equations. This compromise further had the effect of appeasing both sides of the particle/wave debate. Nevertheless, light's newly defined dual character did create an unexpected problem.

As we've already discussed, a powerful feature of Newtonian mechanics was how it could explain the behavior of all objects possessing mass. It could explain everything from the single apple falling from a tree to the majestic motion of the Moon across the night sky. However, Einstein's concept of the photon as a *massless* particle immediately made Newton's formulas no longer applicable to light. So it was now necessary to propose a new, unified equation that could describe the mechanics of both massed and *massless* particles (i.e., energy). And the secret to finding this single equation capable of describing mass and energy was to mathematically derive how they behaved alike and then combine their separate equations.[†] And this brought Einstein back to the

---

[*] The term "photon" was not used to describe light until two decades after Einstein published his 1905 paper. Chemist Gilbert N. Lewis coined the term in a letter to the editor of Nature magazine.
(Vol. 118, Part 2, December 18, 1926, pages 874-875)

[†] Einstein spent some ten years pondering the different aspects of special relativity. However, it must be admitted that it's not certain exactly when and how he realized the formula describing the relationship between mass and energy ($E = mc^2$). Some have suggested that he worked it out mathematically, whereas others insist it was an educated guess. Thus, the logic we use here may not be the same logic Einstein used, but we can still reach the same conclusion.

work of Heinrich Hertz. In 1905, Einstein explained the photo-electric effect, the discovery made by Hertz 18 years prior. Understanding this effect was the key to equating mass and energy. But to fully appreciate how, let's consider a game of billiards and the law of conservation of momentum.

We have a single shot of one billiard ball into the side pocket. After aiming carefully, we impart a force to the pool cue, which, once in motion, gains a certain momentum. When the pool cue strikes the billiard ball, the cue's momentum is transferred to the ball. Finally, the billiard ball heads into the side pocket. (Figure 6-3) And yet, this is not the only means of getting the billiard ball into the side pocket. We could do so just as easily without transferring momentum through the pool cue. For example, we could lift the table. (Figure 6-4) The point is, we can accomplish the same task in a totally different way, yet arrive at the same result: the billiard ball traveling the distance *a* and falling into the side pocket. Let's now see how this explains energy equivalence.

Whether as large as a star or as small as an electron, any object possessing mass also has momentum. Therefore, just as with the example of the billiard ball, we can easily imagine someone carefully aiming a free electron at a second electron orbiting an atom. They then impart a large force to the free electron and watch as it collides with the second, transfers its momentum, and breaks it free of the atom. (Figure 6-5a) And we could easily calculate the minimum speed and, hence, the kinetic energy required of the first electron to dislodge the second. But the critical point to remember is that the same result (i.e., dislodging an electron from its orbit about an atom) is possible using the photoelectric effect.[*] (Figure 6-5b) So although the methods used to dislodge the electrons in Figures 6-5a and 6-5b are entirely different (the first via momentum and the second via energy transfer), the results achieved are identical. And because the results are the same, we can mathematically equate them.

---

[*] The photoelectric effect is the principle upon which solar panels operate to produce electricity. When light from the Sun strikes an electron in a solar panel, the photon's energy transfers to the electron. The electron, now having the extra energy, can break free from the atom. This process creates an electrical current and thus produces electricity.

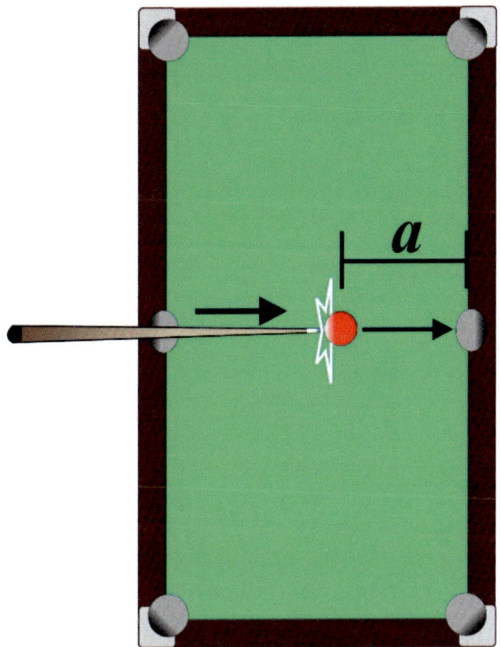

Figure 6-3. To get the billiard ball into the side pocket, energy from the arm of the player creates a force that then imparts momentum to the pool cue. Next, when the cue strikes the billiard ball, the momentum from the cue transfers to the ball, which causes the ball to head for and, finally, enter the side pocket.

Figure 6-4. In this instance, the billiard ball is directed into the side pocket without transferring momentum through the pool cue. Still, the result is the same as in Figure 6-3.

163

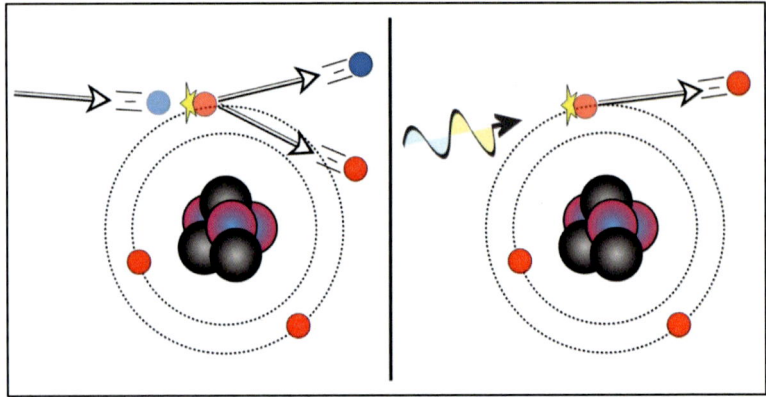

Figure 6-5a (left). An electron with enough momentum can transfer its *momentum* to a second electron, dislodging the second electron from its orbit around the atom. In **Figure 6-5b (right)**, a photon having the correct wavelength and frequency can transfer its *energy* to an electron, giving the electron the strength to break free of its atom. The crucial point is that the same result was achieved using different means, so the two events are comparable.

WARNING: SOME MATH TO FOLLOW

Okay, the above warning may be a bit exaggerated. However, using the example of the photoelectric effect and just a bit of algebra, we can reach the same conclusion as Einstein. We've found a situation where a particle and a photon can accomplish the same task: ejecting an electron from its orbit around an atom. And because we're already quite familiar with momentum, we will say that the photon possesses a momentum-*equivalence*.[*] The next step in comparing the two events is identifying their respective formulas. We'll be comparing momentum, $\rho = mv$, with the formula used to calculate a photon's energy, called Planck's

---

[*] Remember, since a photon of light possesses no mass, it cannot have momentum in the Newtonian sense of the word. However, a photon can eject an electron from an atom via the photoelectric effect. Therefore, the energy the photon used to do so must be equivalent to the momentum of the colliding electron. This specific amount of energy is the photon's momentum-*equivalence*.

equation, $E = hc/\lambda$. Our goal is to make these two formulas equal and thus show how mass and energy are related.[*]

$$\rho = m\text{v} \qquad\qquad E = hc/\lambda$$

In the formula on the left, mass is measured in kilograms (kg), and velocity (v) is measured in meters per second (m/s). In the formula on the right, Planck's constant (h) has the dimensions of kilograms (kg) multiplied by meters squared (m^2) divided by a second (s). And wavelength is measured in meters (m). So, the first thing we must do is rewrite both formulas using their base dimensions. Doing so gives us:

$$\rho = \text{kg} \cdot \text{m/s} \qquad\qquad E = [(\text{kg} \cdot m^2)/\text{s}] \cdot c \cdot 1/\text{m}$$

The next step is to notice that the variable representing the speed of light, c, is not found in the formula for momentum. So, we can move this value to the opposite side in the formula on the right. We do so by dividing both sides by c.

$$\rho = \text{kg} \cdot \text{m/s} \qquad\qquad E/c = [(\text{kg} \cdot m^2)/\text{s}] \cdot 1/\text{m}$$

Still examining the formula for energy, we notice a meters-squared (m^2) and a *reciprocal* meter ($1/\text{m}$) on the right side. When combined, one of the meters in the meters-squared can be removed, $m^2/m = m$. Next, we can remove the parentheses in the equation for energy. We can immediately notice that the two equations are now the same on their right sides.

$$\rho = \text{kg} \cdot \text{m/s} \qquad\qquad E/c = \text{kg} \cdot \text{m/s}$$

But if their right sides are equal, it must also be true that their left sides are as well, which means we can combine them as follows:

$$E/c = \rho$$

---

[*] A more thorough algebraic derivation can be found in *The Einstein Decade (1905-1915)* by Cornelius Lanczos, pages 78-88.

The above formula tells us that the energy contained in a photon (E), divided by the speed of light (c), equals the photon's momentum-equivalence (ρ). Yet, recall from our original equation that momentum (ρ) equals mass (*m*) multiplied by velocity (v). Consequently, we can rewrite the right side of the last equation, ρ, as it was on the right side in its original equation.

$$E/c = mv$$

The only velocity that Einstein was concerned with was the speed of light. We can, therefore, make the velocity (v) on the right side of the equation equal to the speed of light (c), which gives us:

$$E/c = mc$$

Finally, we move the constant, c, on the left side of the equation, back over to the right side by multiplication. When we do so, we arrive at Einstein's famous equation.

$$E = mc \cdot c$$

or

$$E = mc^2$$

The above equation is the most recognized form for mass-energy equivalence. Its formulation was a milestone in scientific history. However, most do not know that this is not the only form of Einstein's famous equation.

The above is a general form of Einstein's mass-energy equation; specifically, it tells us an object's <u>rest mass</u>. However, according to special relativity, an object's relative velocity also contributes to its mass-energy equivalence. In other words, the greater the velocity of an object with mass, the greater its mass-energy equivalence. Therefore, when we factor in the latter, the standard form of Einstein's equation changes to:

$$m = m_0 / \sqrt{1 - v^2/c^2}$$

In this equation, m_0 is the object's rest mass ($E = mc^2$), and m is the object's total *relativistic* mass. And this final form of Einstein's famous equation reveals a powerful truth regarding the very nature of reality.

On the right side of the equation above, notice that if the object's velocity were ever to equal the speed of light, v^2/c^2 becomes c^2/c^2, which equals 1; thus, the denominator equals zero.

$$m = m_0 / \sqrt{1 - 1}$$

or

$$m = m_0 / 0$$

However, one of the **absolute** rules in mathematics is that division by zero is *never* allowed; try it yourself on any calculator. Consequently, this second equation reveals that no object possessing mass can ever reach the speed of light because, if it did, it would result in division by zero. But what does this mean in actual practice?

As an object having mass approaches the speed of light, the equation's denominator begins to shrink as it approaches zero. While on the left side of the equation, the answer grows larger. In effect, the object's mass is becoming exponentially larger. This infinite increase in mass prevents the object from reaching the speed of light. At best, it could *approach* (get close to) the speed of light (i.e., 99.$\overline{999}$%), but it could never achieve it. This conclusion was an essential change from Newtonian mechanics because now all objects had a maximum speed limit. As seen in this formula, special relativity limits the speed of all objects having mass to less than the speed of light. And this limitation has been proven experimentally using particle accelerators. (Figure 6-6a & b)

Scientists use magnets and electric fields in particle accelerators to accelerate subatomic particles in opposing directions. Once these particles reach relativistic speeds, they are collided to expose their constituent parts. Scientists have found that a proton accelerated to 30% of light speed will have a relative mass of 0.05% of its rest mass. However, when it reaches 90% of light speed, its mass increases to 250% of its rest mass. While at

Figure 6-6a. The Tevatron particle accelerator at Fermilab, located in Illinois, USA. Image credit: Fermilab.

Figure 6-6b. A portion of the Tevatron's ring of superconducting magnets. They are used to accelerate particles to up to 99% the speed of light. Image credit: Reider Hahn/Fermilab.

99% of light speed, the proton's relative mass will have increased to 2700% of its rest mass. Notice that this relative increase in mass is not linear. * As a result, no matter the amount of energy put into overcoming that final 1%, the proton's exponential increase in relative mass will prevent it from reaching the speed of light. (Figure 6-7) Yet, there was another reason that Einstein's mass-energy equation was so significant.

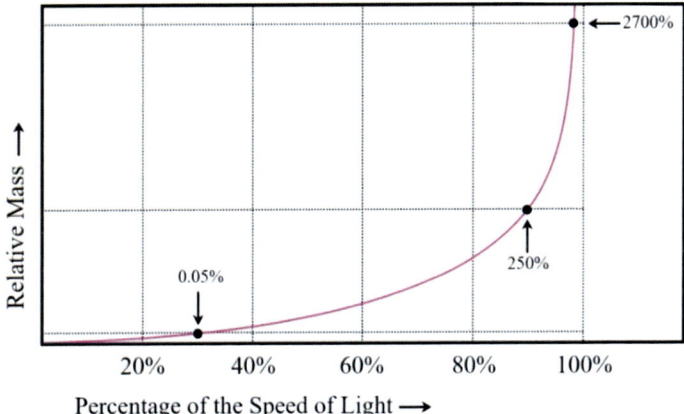

Percentage of the Speed of Light →

Figure 6-7. As a proton is accelerated toward the speed of light, its relative mass increases exponentially. This infinite increase in relative mass prevents the proton from ever reaching light speed.

Beyond uncovering nature's laws, another goal of science is to know how these laws interact. Because, by understanding the interaction of seemingly unrelated laws, we better understand how the universe operates. For example, recall how excited Oersted was when he saw the compass needle deflect. Everyone prior to him had believed that electricity and magnetism were unrelated. But after that day more than 200 years ago, we no longer spoke of those two forces separately; from that moment on, they were forever united as electromagnetism. And once it was known how those forces interacted, our ability to utilize them became possible;

---

* To be clear, the literal mass of the proton has not increased. Instead, its momentum is growing. At 99% of the speed of light, the proton's momentum is equivalent to that of 2,700 protons!

indeed, it's the fundamental basis of electric power generation. The same can be said of Einstein's mass-energy equation.

Before Einstein, energy and matter were considered separate and distinct. But with special relativity, they were now seen as two sides of the same coin. All mass contains a vast amount of energy; all energy has a momentum-equivalence. Grasping this relationship at once allowed scientists to imagine many new possibilities. They could conceive of producing energy by splitting the atom – turning heavier elements into lighter ones in the process known as fission. At the same time, others saw how the Sun reversed that process – releasing vast amounts of energy by fusing lighter elements into heavier ones – through nuclear fusion. How much energy is produced in these processes? $E = mc^2$. And yet, as incredible as Einstein's achievements were, special relativity was only a partial solution.

As its name implies, special relativity only applies to the *special* case of objects under *uniform* velocity. But here on Earth, uniform velocity is not the norm. Instead, we, along with everything around us, are in a constant state of change. Even a plane, flying at a steady speed in one direction, is actually accelerating because, to maintain its altitude, it must continually follow the *curvature* of the Earth.* So the next question was: how does relativity apply to objects that are accelerating? Or, more specifically: What exactly is gravity? This question was what *general* relativity would seek to explain.

---

* Uniform velocity means that both speed and direction are constant. An example of uniform velocity would be traveling at 30 km/hr in a straight line without speeding up, slowing down, or changing direction. Acceleration, on the other hand, means you are speeding up, slowing down, or changing direction. For instance, a car can make a right turn without "slowing down." But since it is changing its direction of travel as it turns, it is decelerating in its previous direction and accelerating in its new direction. Consequently, its velocity is changing. Changing speed or direction, such as traveling in a circle, is a change in velocity; your velocity would no longer be uniform (constant).

Albert Einstein
(from 1905 to 1915)

After his miracle year, Einstein's recognition was not immediate, primarily because the concepts of special relativity and his description of the nature of light were unlike anything scientists had ever seen. As a result, most didn't know how to respond. For example, before 1905, another of history's great scientific minds, Max Planck, had been trying to reconcile how light energy always occurred in definite packets, which he called *quanta.*[*] Planck had not been able to explain this phenomenon. Yet when Einstein published

**Max Planck
(1858-1947)**

his revolutionary 1905 paper, in which he described the photon, the first person to reject the explanation (although it answered the very question he sought) was Max Planck. Planck reasoned that to accept Einstein's interpretation of the photon, it behaves as both a particle and a wave, meant to reject Maxwell's Equations. However, Planck didn't totally reject Einstein's theories.

Most in the scientific community found special relativity confusing and couldn't grasp its subtle beauty. Nevertheless, Planck, who initially found parts of the theory hard to conceive, gradually accepted it. In fact, if it were not for Planck's support and ardent endorsement of special relativity, chances are it would've taken even longer for the theory to gain acceptance. And yet, in the short term, Einstein's circumstances didn't change. Although awarded his Ph.D. in physics in January 1906,[†] he could

---

[*] In physics, a quantum (plural: quanta) is a discrete *quantity* of energy. For example, a photon of ordinary white light is composed of a <u>fixed amount</u> of energy. This quantum of energy is the smallest, *indivisible* amount of energy required to make up that single photon of white light.

[†] Ironically, Einstein's Ph.D. thesis, though summited the same year, is not usually counted among the revolutionary papers of his miracle year of 1905. However, it did form the basis of another of his celebrated papers, that of Brownian motion. His thesis was entitled *A New Determination of Molecular Dimensions*.

still find no work as a professor. Thus, he was forced to remain at the Swiss patent office for another year. But things finally began to change in 1907.

**Johannes Stark
(1874-1957)**

A year after Einstein earned his Ph.D., the scientific community finally began appreciating special relativity. It was around this time that German physicist Johannes Stark* requested that Einstein submit an article for the journal *Jahrbuch der Radioaktivitaet und Elektronik*. Stark wanted Einstein to explain how his theory of special relativity related to the then-current understanding of physics. As it turned out, this request is what started Einstein on the path toward a non-special (or *general*) theory of relativity. Because, as Einstein wrote the article, he found it easy to resolve special relativity with electrodynamics since it was based on Maxwell's Equations. However, he quickly realized that special relativity could not be reconciled to Newtonian gravity. Therefore, while he worked full-time at the patent office, sought employment as a professor, and completed his article for Stark, Einstein also began to work on a more all-purpose form of relativity. This general form of the theory would incorporate and, even more importantly, attempt to explain once and for all exactly what gravity was. To say that Einstein had a full schedule would be an understatement.

In 1908, the University of Bern accepted Einstein's academic credentials; he gave his first address as an unpaid guest lecturer by the end of that year. But sadly, just as his scientific career was beginning to flourish, Einstein's personal life began to fall apart. Especially beginning with his "miracle year," did his marriage start to suffer. Einstein was paying more attention to his work than to his family. And the more his creativity surged, the more neglectful he became, so much so that Mileva began to suffer from severe depression. To make things worse, only revealed in 2006 by letters from his now-deceased stepdaughter, Einstein was also a chronic philanderer.

---

* Image Credit: Nobelprize.org

**Elsa Einstein
(1876-1936)**

It is estimated that between the time he married his first wife and the death of his second, Einstein may have had as many as a dozen mistresses. Further, it appears that he didn't even bother to hide these offenses from either of his wives. Of course, Mileva was very disturbed by it all, which undoubtedly was the cause of her depression. She eventually suffered a nervous breakdown upon their divorce in 1919. Einstein's second wife was his cousin, Elsa.* Their affair began in 1912 while he was on a trip to Berlin. Two years later, Einstein left his family to be with her. Then, in 1919, immediately upon his divorcing Mileva, Einstein married Elsa.

It appears that Einstein's and Elsa's almost 20-year marriage may have been nothing more than platonic, at best. Not long after they wed, Einstein began an affair with Betty Neumann, a friend's niece. But while Einstein's conduct distressed Mileva, Elsa openly tolerated it. Some speculate she accepted the first affair to prevent him from finding even more mistresses. Others say she only married Einstein because of the popularity she gained. Einstein, on the other hand, blatantly stated that he married Elsa only because she cooked well. It was amid this family dysfunction – after more than five years of work on general relativity – that Einstein was finally driven to a sober realization. He realized that he needed the help of a mathematician.

Modern courses explaining special relativity have as a prerequisite algebra and first-semester calculus. But not so with general relativity. The latter theory relies heavily on differential geometry and tensor calculus. And although Einstein understood advanced mathematics, he was still only a physicist, not a mathematician; however, he did know someone who could help him.

---

* Image credit: German Federal Archive

**Marcel Grossmann
(1878-1936)**

While at the Swiss Federal Institute, Einstein had become close to a man named Marcel Grossmann.[*] Indeed, only with Grossmann's help did Einstein graduate with his first degree. As Einstein skipped many of his lectures, Grossmann provided notes for Einstein to review. So it was not out of character that, in 1912, Einstein enlisted his friend, now a Doctor of Mathematics. The result of this collaboration was that, when Einstein finally revealed his finished paper on general relativity, its technical *and* mathematical beauty was unmatched. Indeed, an argument can even be made that Grossmann's math is what helped the scientific community to accept general relativity more quickly than special relativity.[†]

Einstein and Grossmann's first draft on general relativity in 1913 comprised two sections. The first section was the physics of relativity, written by Einstein. Grossmann composed the second section, which was the mathematics of relativity. And finally, in 1915, Einstein's general theory of relativity was ready to be presented. However, to understand how he formulated his general theory of relativity, let's begin by examining a major problem Einstein had with Newtonian mechanics and gravity.

---

[*] Image credit: http://learn-math.info/

[†] In honor of Grossman's contributions to general relativity – beginning in 1975 and every three years after – an annual conference is held: the Marcel Grossmann Meetings. At these meetings, any mathematical discoveries in general or relativistic field theory are presented.

Chapter 7:
The General Theory of Relativity

The discoveries that contributed to the development of special relativity were like many stray fragments waiting to be assembled. However, the physics of general relativity was the sole conception of Albert Einstein. No one before him had ever conceived of gravity in the fashion he did and for a good reason. The challenge in studying gravity is that nothing in nature behaves quite like it. For example, most of the forces we know of are dichotomous. Electricity can be positive or negative, magnetism has a north and a south pole, and even forces that arise due to motion can be positive or negative depending on the direction of application. The ability to examine a force by comparing it to a counterpart provides insight into what is being studied. Think again of Oersted's discovery that magnetism and electricity were related.

If Oersted had never seen the connection between electricity and magnetism, there would be no electric power generation. We have electric power because a current moving through a wire produces a magnetic field and vice versa. So, whether using water pressure behind a dam, heating water to create steam, or using wind, they all share one thing in common. They all involve a turbine rotating a coil of wire between magnets. And each time the coil passes the magnet, an electrical current is induced in the wires. Electric production exists because Oersted discovered that electricity and magnetism were counterparts. But what opposite (or counterpart) force is there to gravity? If there is one, we have yet to find it. Therefore, no one has ever seen gravity behave in any other way than as an attractive force, which makes the study of gravity extremely difficult. It's almost as if gravity stands alone among the four fundamental forces. So, how was Einstein able to overcome this drawback to deduce general relativity and, in the process, define the nature of gravity? He began by analyzing the flaws found in Newton's description of gravity.

Another Problem with Newtonian Mechanics

Einstein's main objection with Newtonian physics was action-at-a-distance, the idea that gravity exerted an *instantaneously* force over infinite distances. To counter this idea, Einstein proposed relativity of simultaneity. This latter concept proposed that the information that signifies an event as having occurred is also of a physical nature and, thus, is subject to physical laws. And, since all physical phenomena require time to propagate, this placed a speed limit on how fast information could travel, including gravity. So, relativity of simultaneity does not allow gravity to be instantaneous; gravity does not propagate faster than the speed of light. This solution resolved Einstein's first issue with Newtonian physics. But a second, more significant issue that he had with Newtonian gravity was that, upon closer examination, it didn't obey the principle of relativity. Consider again the experiment Galileo conducted atop the Tower of Pisa to understand why.

Galileo wanted to prove that all objects fell at the same velocity, so he dropped both a heavy and heavier object from the Tower of Pisa. And each time he conducted the experiment, the results were the same: both objects struck the ground simultaneously, no matter their mass. Now, imagine conducting Galileo's experiment again, but this time with a cannon. Furthermore, we'll run the experiment using Newton's laws of motion and explanation of gravity.

You are atop the Tower of Pisa and have two identical cannonballs. As in the previous thought experiments with the apple, the first cannonball we will drop straight down. However, the second cannonball we will fire horizontally from a cannon, propelling it to half the speed of light. Here's the question: Will they fall at the same rate toward the ground? We drop the first cannonball at the same instance that we fire the second horizontally.

Now, the force gravity exerts on both cannonballs is the same, 9.8 m/s^2. Newtonian mechanics wants us to think that the cannonballs will reach the ground simultaneously. However, it must be remembered that the inertia of the two cannonballs is no longer equal. And that presents a significant problem.

Inertia is defined as the tendency of an object to *resist* acceleration. The cannonball moving at half the speed of light has greater inertia than the dropped one. And since gravity causes all objects to *accelerate* toward the ground, the second cannonball's greater inertia would also mean it has a greater resistance to gravity. Or, in other words, the second cannonball should fall more slowly than the first. (Figure 7-1) And this unsuspected consequence contradicted special relativity.

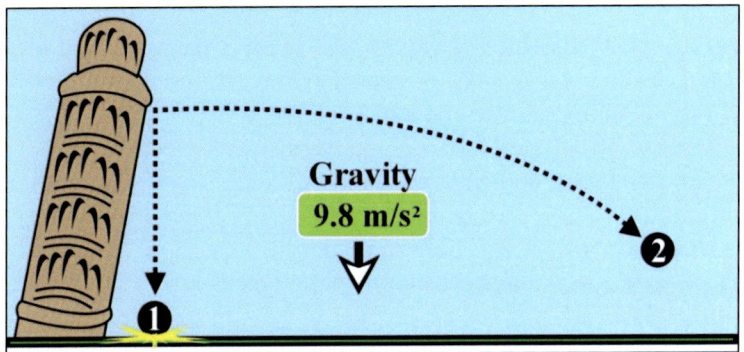

Figure 7-1. The cannonball possessing the horizontal momentum will also possess more inertia; thus, it will have a greater resistance to gravity. The result would be that, according to Newtonian mechanics, the second cannonball will fall slower than the first, even though both are experiencing the same gravitational force.

Recall that, according to special relativity, increased velocity is equated with an increase in energy. At the same time, increased energy is linked with increased mass (i.e., $E = mc^2$). Hence, when Newton's law of inertia is combined with special relativity, it implies that objects with more mass (or energy) <u>fall slower</u> than lighter (or less energetic) objects. So even if the second cannonball was merely heated to a higher temperature than the first, it should still fall more slowly. This result made it clear to Einstein that Newtonian gravity could not be reconciled with special relativity. And he was not alone in realizing the emerging paradoxes in Newtonian physics.

By the start of the 20th century, other scientists were also becoming aware of discrepancies within Newtonian physics. But, at the time, these inconsistencies were easy to ignore because, at everyday velocities and distances, they were negligible, which remains true even today. However, the only velocity that Einstein was obsessed with was the speed of light. And when dealing with these extreme velocities, inertia and Newtonian gravity could not be reconciled with either Galilean or special relativity. Accordingly, Einstein's next challenge became clear. He needed to do what no one else had done; he needed to define the nature of gravity.

As it turned out, Einstein would present a solution that was not only original but was surprisingly straightforward. Instead of continuing to imagine that gravity was a *force* that emanated from objects having mass (like magnetism), he proposed that gravity was the bending or distortion of space-time itself. Large objects, like the Earth, distorted space-time significantly, causing smaller objects to accelerate toward the center of that distorted space-time. This simple solution instantly resolved the problems that plagued Newtonian physics. But how did Einstein reach this conclusion? By following the same procedure that he had used to formulate his special relativity theory: he devised a thought experiment, came up with a principle, and then examined its consequences.

The Equivalence Principle

While looking out the patent office window one day, Einstein watched laborers working on an adjacent roof. As he did so, he asked himself the question: What would happen if one of the laborers fell from the rooftop with his tools? Well, according to Galilean relativity, all objects within the same gravitational field fall at the same rate. Therefore, as the laborer fell from the roof, any tools that fell with him would fall at the same rate. As a result, they (the laborer and the tools) would be <u>motionless relative to one another</u>. In other words, objects in freefall behave – relative to one another – as if they are floating motionlessly in space. And this was Einstein's "Eureka!" moment. Here was a situation where objects <u>in a gravitational field behaved as if they were not</u>. And, consequently, the principles of special relativity could be applied. Stated another way: a freefalling object in a gravitational field can be treated the same as an object moving at uniform velocity in the absence of gravity. This insight was the single, most critical moment of Einstein's conception of gravity and the formulation of general relativity. But, to understand why, put yourself in the position of the falling laborer. However, instead of falling from a rooftop, imagine falling toward the Moon, inside a large box with no windows.

Suppose for several weeks you've floated weightlessly in space inside a large box. (Figure 7-2) Now, unknown to you, you've just come close enough to the Moon to start to be affected by its gravity. So, here's the question: Would you know the exact moment the Moon's gravity began to attract you? The answer is no. As long as you *and* the box are accelerated *at the same rate*, you *and* the box will be stationary relative to one another. And since the box has no windows, you wouldn't see the fast-approaching lunar surface. Plus, the moon has no atmosphere to

Figure 7-2. Einstein realized that, just as a person in uniform motion (or at rest) in the middle of space would be weightless, so too a person being uniformly accelerated. Einstein, therefore, concluded that uniform acceleration, just as with uniform velocity, was relative.

slow the box, so your feeling of weightlessness would remain until the moment of impact. You would have no way of knowing that you were in freefall. Indeed, this is the same reason that the astronauts onboard the international space station are weightless.

Like the Moon, the orbiting astronauts are held by the Earth's gravity. And the fact that the station doesn't fly off into the depths of space is evidence that it too is within the Earth's gravity. Yet, the astronauts onboard the station experience weightlessness because they *and* the space station are orbiting (falling) at identical rates. In other words, although held within the Earth's gravity, the astronauts don't "feel" the Earth's gravity pulling on them. Einstein realized that freefall (or falling with a uniform acceleration) was equivalent to uniform velocity. Now, what consequence did this fact imply?

Recall Galileo's ship. No experiment can be conducted while traveling at a uniform velocity that could tell you whether you were at rest or not; this is the principle of relativity. Einstein now expanded this: no experiment can be conducted in a uniformly *accelerating* frame of reference that is capable of telling you whether you were accelerating or not. Consequently, as experienced during freefall, uniform acceleration is also relative, whether you are uniformly accelerating at 10 000 m/s^2 or 9.8 m/s^2. As long as the acceleration is uniform, no experiment could tell you that you were in freefall. Thus, if you jumped from a plane and you couldn't see the approaching ground, and there was no air blowing past, you wouldn't know that you were falling.

Therefore, having made the connection between uniform velocity and uniform acceleration (i.e., freefall in a gravitational field), Einstein was now overturning Newtonian gravity. However, this was only the first step. Next, he would have to explain how the principle of relativity applied to a person in a gravitational field who was *not* in freefall, which was a more complicated question. Still, as he had done with special relativity, Einstein employed another thought experiment: that of the elevator.

In the last example, we considered acceleration (freefall) within a gravitational field; gravity accelerated both the box *and* the box's occupant equally. Consequently, the box and the person in the box remained motionless relative to one another. Now suppose that a person was inside an elevator in space. If there were no nearby planet, thus no gravity, he and the elevator would be weightless. But what would happen if *only* the elevator began to accelerate at a rate of 9.8 m/s^2? (Figure 7-3)

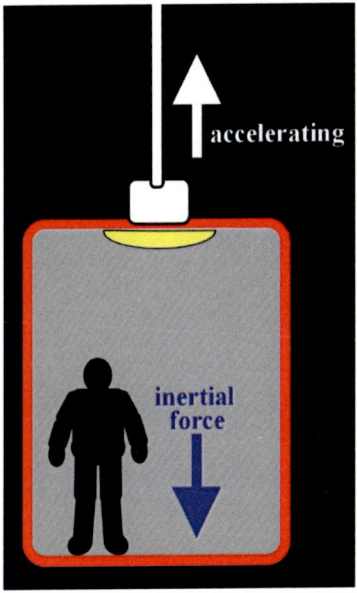

Figure 7-3. Einstein deduced that, outside a gravitational field, a person in an accelerating elevator would feel as if he were standing motionless on the Earth's surface. There is no difference between the acceleration caused by the elevator and that caused by gravity.

Since the elevator is being accelerated and *not* the person inside the elevator, the person's inertia will resist the acceleration once he makes contact with the elevator floor. Subsequently, as the elevator continues to accelerate (the arrow pointing up in Figure 7-3), the person's inertia will resist in the opposite direction (the arrow pointing down in Figure 7-3). Yet, what makes this thought experiment interesting is that the elevator's rate of acceleration is the same that's found at Earth's surface. Therefore, the person in the elevator will *feel* just as he would if he were standing motionless on the Earth's surface. Even if he stood on a scale to weigh himself in the elevator, he would register the same weight that he would on Earth's surface. And since the elevator has no windows, he would have no way of knowing if he was in a motionless elevator on the Earth's surface or in an accelerating elevator in space. Indeed, to further demonstrate how the latter two situations are equivalent, let's consider an experiment. If you released a ball on the Earth's surface, it would fall toward the ground. At what speed will it fall? At 9.8 m/s^2. We will now conduct the same experiment in the above elevator in space under equal acceleration. (Figure 7-4) What would be the results?

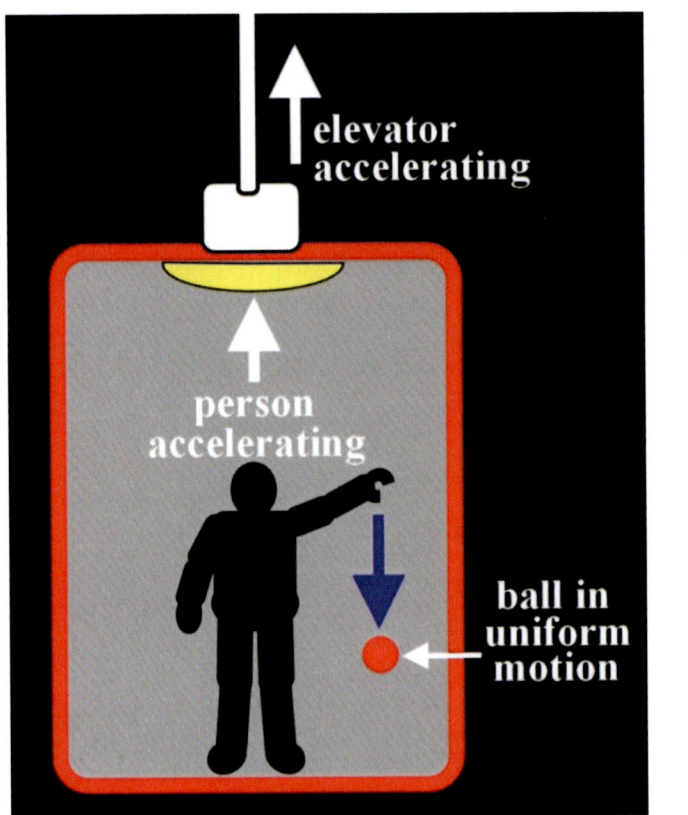

Figure 7-4. Once the ball is released, it is no longer being accelerated. Instead, it's now traveling forward in the same direction as the elevator but with a *uniform* velocity. However, because the elevator is still *accelerating*, the ball will *appear* to fall to the elevator floor relative to the person inside.

Newton says that an object in motion will remain in motion so long as no other forces act upon it. So, whatever the elevator's speed at the instant the ball is released is the speed at which the ball will continue to move forward. Let's say the elevator is accelerating at 9.8 m/s^2 and has been accelerating for 1 second; it has just reached a speed of 9.8 m/s. And this is the moment the person (who is still in contact with the accelerating elevator) releases the ball. Once released, the ball will no longer be accelerating. However, given Newton's 1st law of motion stated above, the ball would continue to move *forward in the same direction* but at the uniform (non-accelerating) velocity of 9.8 m/s (no seconds squared). So, how will the ball *appear* to behave to the person in the elevator?

Once released, the ball travels with a uniform velocity in the same direction as the elevator. However, the elevator and the person inside are still *accelerating*. As a result, after being released, the ball will *appear* to fall to the elevator floor immediately. Now, in reality, it is the elevator floor that's *accelerating* toward the ball. (Figure 7-4) At what speed is the elevator accelerating? At 9.8 m/s^2. And given that the elevator's acceleration matches that caused by Earth's gravity, the ball will *appear* to behave just as it would if you dropped it while standing on the Earth's surface. It would even bounce when it strikes the elevator floor. Therefore, by applying the *duck test,*[*] Einstein concluded that gravity was not like magnetism; gravity was not a *force*. And while this thought experiment may seem simple, it changed the nature of science forever. Consider.

An *accelerating* elevator can simulate the effects of gravity; therefore, the ball above *appears* to fall. On the other hand, gravity causes the same ball to accelerate (i.e., fall) on the Earth's surface. Gravity causes mass to accelerate, while acceleration simulates gravitational effects on objects with mass. In effect, Einstein had found a phenomenon (acceleration) to contrast with gravity. And it was this contrast that allowed him to conclude that gravity *was not a force* like magnetism.

---

[*] The duck test is attributed to American writer and poet James Whitcomb Riley. He wrote in one of his poems: "When I see a bird that walks like a duck and swims like a duck and quacks like a duck, I call that bird a duck."

Gravity causes objects with mass to accelerate (i.e., fall). But what is acceleration? Acceleration is defined as a continuous increase in *speed*. And what is speed? Distance (*space*) divided by *time*. Therefore, just as the acceleration of an object is its movement through *space* over *time*, Einstein concluded that gravity must be doing the opposite; gravity distorts *space* and *time* to accelerate objects. And this simple thought experiment of an elevator accelerating in space allowed Einstein to now postulate the *equivalence principle*.

The Equivalence Principle:
> The effects of a gravitational field are indistinguishable from the effects caused by an equivalent rate of acceleration.[*]

The equivalence principle was the most innovative conception of gravity since Isaac Newton. However, once armed with this principle, Einstein needed to prove it experimentally. How could he verify that gravity causes objects to accelerate by manipulating *space* and *time*? Let's first return to the elevator thought experiment to consider the consequences of his theory.

Effects on Light, Space, and Time

Imagine the same elevator as in the thought experiment above, except now you're standing outside the elevator, motionless. You also share the same inertial frame of reference as a light source emitting a beam of light. (Figure 7-5) You watch as the light beam enters the moving elevator from the left, 3 meters above the elevator floor, and as it exits the other side a fraction of

---

[*] Just to be clear, acceleration and gravity are not the same things. When a person is accelerating, such as in a vehicle, he is in relative motion. But this is not the case with gravity, such as when standing relatively motionless on the Earth's surface. The effects, however, are the same. A person can feel his inertia resisting as he's pressed against the seat in an accelerating vehicle; he can feel himself getting heavier. Similarly, when standing on the Earth's surface, you can feel your inertia resist the Earth's gravity; that feeling of weight is what keeps you securely planted on the ground.

a second later. Since you are motionless relative to the light source, then relative to you, the beam of light maintains a straight, horizontal path. Thus, the only thing that appears to change position is the elevator. Let's assume it traveled 0.9 meters as the light passed through it. As described, the light beam would appear as pictured in Figure 7-5. Now, let's examine the same situation, but this time from <u>inside</u> the elevator.

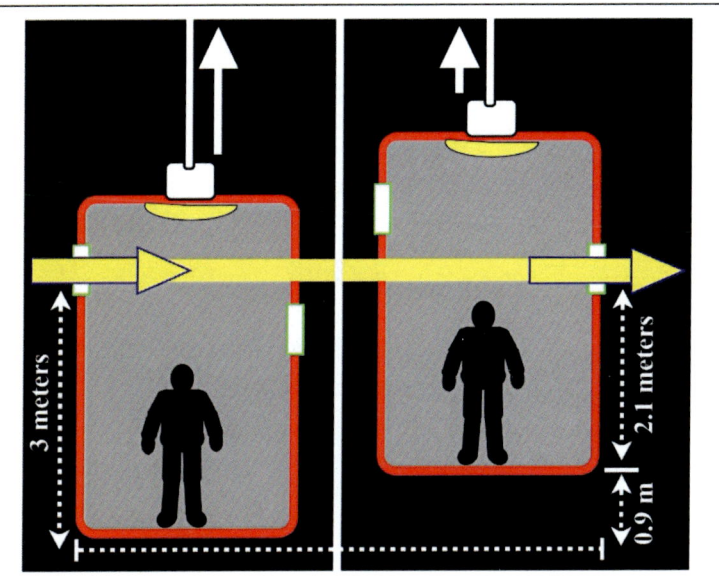

Figure 7-5. When viewed from *outside* the elevator, standing relatively *motionless*, the beam of light travels in a straight line. In contrast, the elevator appears to be in motion. Therefore, the light enters the elevator 3 meters above its floor, only to exit after the elevator has traveled 0.9 meters; the beam of light exits the elevator 2.1 meters above the floor.

Now imagine you're *inside* the elevator, and it's moving at a *uniform* velocity. (Figure 7-6) According to Newton's 1st law, uniform speed is equivalent to a state of rest; in other words, you wouldn't *feel* as if you're moving. However, since you are moving relative to the light source, you will observe the light differently than a person outside the elevator. You would still see the light enter the elevator on your right, 3 meters above the floor, and exit the elevator on your left, 2.1 meters above the floor. But, as with stellar aberration, since you and the elevator are traveling at a *uniform* velocity, the light will appear to travel at a downward angle. So, to the person <u>outside</u> the elevator, who is relatively motionless, the beam of light moves horizontally through the elevator. But to you, since you are in relative (uniform) motion, the beam of light will appear to cross the elevator at a <u>downward</u> angle, as in Figure 7-6. Just as predicted by Galilean relativity, based on the different inertial frames of reference, the path of the beam of light will appear different.

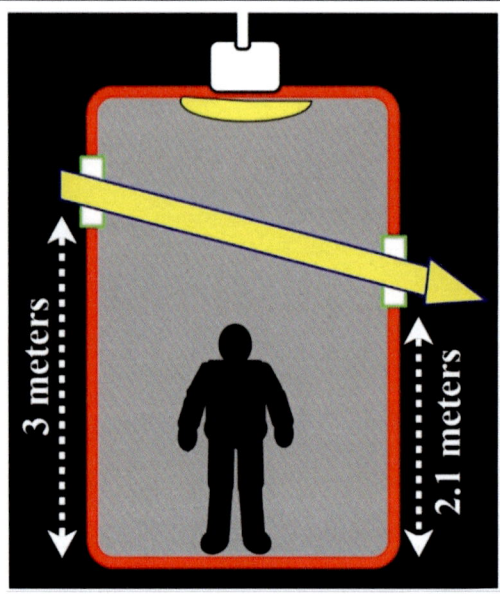

Figure 7-6. According to Newton's 1st law, someone inside a *uniformly* moving elevator is *relatively* motionless. However, his motion relative to the light source causes the light beam to appear to cross the elevator along a straight path but at an angle.

However, in Einstein's thought experiment, the elevator wasn't moving at a *uniform* velocity. Instead, simulating the Earth's gravity, it was uniformly *accelerating*. So, how would the beam of light appear to move to someone *inside* an accelerating elevator? The quick answer is that it would appear to *curve* downward. (Figure 7-7) To understand why let's consider Einstein's thought experiment a little more closely.

186

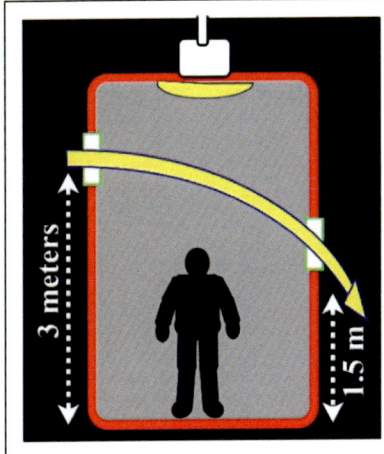

Figure 7-7. Relative to a person inside an elevator moving with uniform acceleration, the beam of light will appear to curve as it passes through the elevator.

Let's assume the elevator in Figure 7-8 is uniformly *accelerating* at a rate of 1 meter per second squared (1 m/s^2). The elevator's speed increases by 1 meter per second for every second of travel. The numbers on the left represent the elevator's position as measured from its starting point. The column on the right counts the number of seconds the elevator has traveled. So, begin at position 0, let's track the light's movement through the elevator.

Starting at the bottom in Figure 7-8, we know that the light (the yellow photon) supplies the horizontal motion (left to right), and its speed is constant. On the other hand, the upward movement of the elevator supplies the relative downward motion of the photon, just as it did in Figure 7-6. However, unlike the constant velocity of the elevator in Figure 7-6, which causes the light to have a <u>set</u> angle downward, the elevator now *accelerates*. In other words, the elevator *increases* its speed after each second of travel. Correspondingly, then, after each second of travel, the downward angle of the photon will also appear to *increase*. Thus, as you track the photon at each consecutive second (each box), the combination of the constant horizontal motion with an *accelerating* frame of reference causes the photon to follow a seemingly curved trajectory. (Figure 7-8) This is the same phenomenon we saw in Galileo's ship experiment. The ship's motion combined with gravity caused the droplet to appear to curve as it fell. (Figure 2-7b) Or with Newton's apple following a curved trajectory when thrown horizontally. (Figure 3-2)

187

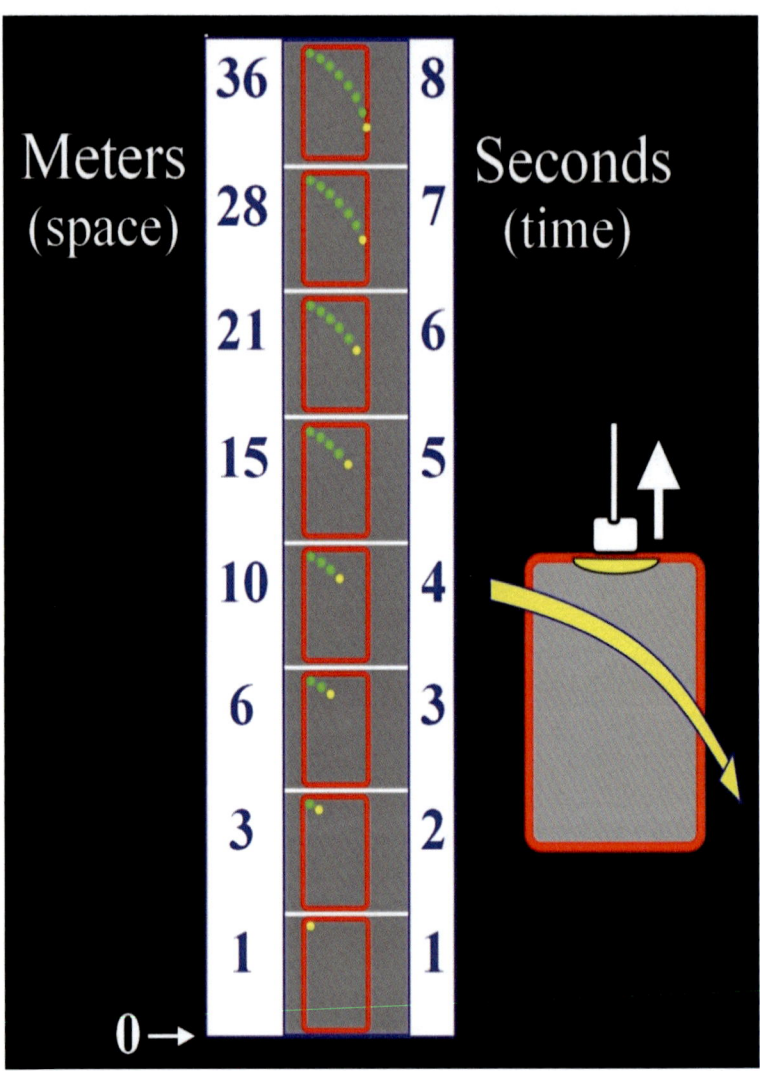

Figure 7-8. For each moment of travel through the elevator, the angle of the photon appears to become steeper. However, the light's constant (straight-line) horizontal speed hasn't changed. From this thought experiment, Einstein concluded that the elevator's continually changing position in <u>space</u> over <u>time</u> was causing the light beam to appear to curve.

Therefore, the constant speed of light, combined with the elevator's acceleration, causes the light beam to *appear* to curve downward. And with this crucial observation in mind, now recall the equivalence principle. "The effects of a gravitational field are indistinguishable from those caused by an equivalent rate of *acceleration*." So the elevator's acceleration will make it appear that light curves. If Einstein were correct, gravity would reverse this process; gravity should distort space and time and thus cause light to curve. By finding an analog to gravity in acceleration, Einstein could *define* gravity as the distortion of space and time.

So, to quickly summarize, Einstein proved that gravity was not a force, as Newton had thought. Because if it were, then an object with more energy (or velocity) would fall slower than the same object with less energy (or velocity). (Figure 7-1) Next, the equivalence principle compared gravity to acceleration. However, since acceleration is defined as changing position in *space* over *time*, the true equivalency was between gravity and space-time. The result of these two conclusions is General Relativity.

> ➢ Gravity is the *distortion* of *space-time* around objects possessing mass, which creates a gravity well into which objects possessing mass are accelerated (or falls). [See Figure 1]

And what makes this conception of gravity so revolutionary is that it solved every one of the problems associated with Newtonian gravity.

For example, since general relativity didn't treat gravity as a force but as a physical property of space, it eliminated Newton's action-at-a-distance. Now, like every other physical phenomenon, gravity requires time to propagate. Furthermore, since gravity was now the literal distortion of space and time, the aether was no longer needed to transmit it. Even more significant, Einstein's theory upheld Galilean relativity. Thus, all objects in the same gravitational field, be they a feather or a hammer, will fall at the same rate. Yes, every deficiency associated with Newtonian gravity was solved by general relativity. All that now remained was for Einstein to make sure that his new theory reconciled with his special theory of relativity.

189

Special relativity is concerned with *uniform* velocities. General relativity deals with *uniform* acceleration. However, Einstein had shown that both uniform velocity and uniform acceleration adhered to Galilean relativity. Two objects traveling at uniform velocity appear motionless to one another. Similarly, two objects accelerating at the same rate will also appear motionless to one another. Hence, uniform velocity in special relativity was easily reconciled with uniform acceleration in general relativity; they produced the same consequences. Next, special relativity suggested that as the velocity of an object approached the speed of light, time slowed down. To keep this consequence valid in general relativity, Einstein proposed that: As the strength of a gravitational field increases, time will slow down. Consider why this *must* be the case.

According to special relativity, as the velocity of an object increases, the measured rate of time near that object decreases. So what's the equivalent process in general relativity? Well, we start with the fact that the farther you are inside a gravitational field, the stronger the force of gravity (i.e., the *accelerating* force of gravity increases).* Therefore, when standing motionless on the Earth's surface, which is farther into Earth's gravity well than on top of Mt. Everest, the increased gravity means you *would* more quickly accelerate if the ground beneath you were to disappear. And moving faster, according to special relativity, equates to the *slowing* of the rate of time. Therefore, gravity *literally* slows time as its strength increases. And this phenomenon leads directly to one of the theory's most bizarre consequences.

Recall that in special relativity, as an object's velocity increases, there is a decrease in that object's length, called *length* contraction. A similar process also occurs in general relativity. All objects having mass in a gravitational field are contracted (or

---

* For instance, at the top of Mt. Everest, which is 8850 meters above the Earth's surface, the acceleration caused by gravity is $9.773 m/s^2$. At sea level, the acceleration caused by gravity is $9.8 m/s^2$. Meanwhile, ignoring the water pressure at the bottom of the Mariana Trench (which is the lowest place on Earth at 11 260 meters below sea level), the acceleration caused by gravity increases to $9.823 m/s^2$. So the farther you are inside a gravitational field, the greater the acceleration caused by gravity.

squeezed) – the greater the gravity, the greater the contraction. And such *conic-shaped* contraction can become so strong that the mass of a star, for example, can be squeezed down to a single infinite point: a black hole.* And one unique feature of this peculiar star would be that, at its event horizon, the force of gravity would increase to such an extent that time would appear to stop. If you were to watch someone fall into a black hole, they would appear to become motionless at the event horizon.

Light, although possessing no mass, curving. The rate of time slowing as the strength of gravity increases. And black holes in space. Yes, these were the consequences of general relativity. Einstein had even defined gravity – it is the distortion of *space-time* itself. And with this final achievement in 1915, all that remained was to devise a way of testing general relativity and showing that it was, indeed, a real-world phenomenon. And Einstein himself was the one to conceive of the experiment.

Proof of Theory

The initial indication that general relativity was correct came not long after Einstein proposed the theory. General relativity predicted that the nearer you were to a massive object, the greater that object distorted space-time. However, in addition to bending space-time, Einstein stated that if the massive object were rotating, it would also cause space-time to *twist*. He called this phenomenon *frame-dragging*. It occurs because rotational *speed* is also defined in terms of distance (*space*) divided by *time*. So, if a massively large object is rotating quickly, its mass would also twist space-time as it did so. And at once, it was realized that this was the solution to Mercury's perihelion shift. We can compare frame-dragging to a leaf in a large tub of water.

Imagine a long tub of water. At the back-end of the tub, the water is calm as it carries a leaf toward the drain. At the front of the tub, the water circles the drain before emptying. As a result, as the leaf nears the drain, it too gets caught in the circling currents,

---

* And keep in mind that it's not that the "force" of gravity is squeezing the object. Gravity is the distortion of space and time around massive objects, which, in turn, causes the object to volumetrically contract.

so it doesn't immediately go down the drain. Instead, the leaf is quickly rotated *around* the drain. Indeed, the leaf is slightly catapulted away, only to be drawn back toward the drain to circle it once more. And as you watch, you notice that the path the leaf takes is different each time it orbits the drain. This circling, then catapulting of the leaf, can occur multiple times before it finally enters the drain. We have probably witnessed the above scenario, or something similar, many times. Yet, the behavior of this leaf mirrors, almost precisely, general relativity's effect on Mercury.

At a far distance from the Sun, its gravity is smooth – objects are drawn toward the Sun along a straight-line path; the shape of space-time within the solar system is flat at large distances from the Sun. But nearer the Sun, a little smaller than the radius of Mercury's orbit, space-time loses that smooth texture as the Sun's rotation also *twists* space-time. * Therefore, each time Mercury reaches its perihelion (its closest point to the Sun), like the leaf near the drain, frame-dragging whips it about slightly. And this slight shift causes Mercury's perihelion to advance just a bit from where it was previously. (Figure 7-9) Mercury's perihelion shift precisely matched what was predicted by general relativity, which was a crucial first proof.

But the first experiment that sought to prove general relativity was correct was suggested by Einstein in his paper *On the Influence of Gravitation on the Propagation of Light*. He recalled his elevator thought experiment; any light passing near a massive object should curve as it does so. Again, the light is not bending because gravity is pulling it – light has no mass for gravity to affect. Instead, since gravity *is the curving of* space-time, any light crossing that curved space is forced to travel along that *curved* path. Therefore, Einstein suggested that it might be possible to see the deflection of starlight as it passed near the most massive object in our solar system – the Sun. However, there was one problem. The Sun is far too bright to observe starlight passing near its edge. So, how would it be possible to observe the faint light from distant

---

* The Sun has a mass of 1.9891×10^{30} kilograms, which is more than 330,000 times greater than the mass of the Earth. The Sun rotates at approximately 7200 kilometers per hour, more than seven times faster than the Earth. The result of this massively rotating object is the twisting of space-time.

stars curve as it passed near the Sun's surface? In a spark of cleverness, Einstein proposed that this might be possible during a solar eclipse. Therefore, if general relativity was correct, while the Moon blocked most of the Sun's light, it should be possible for stars located *behind* the Sun to be seen. And from the moment it was suggested, astronomers from all over the world scrambled to view the next solar eclipse. They all wanted to be the first to verify whether the amount of deflection predicted by Einstein was correct, thereby proving general relativity.

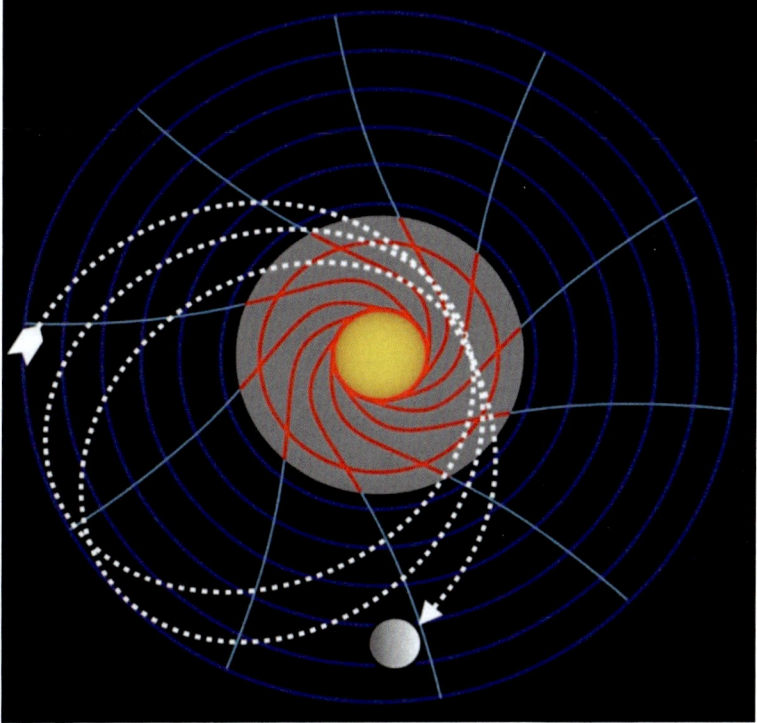

Figure 7-9. The Sun's mass creates a gravity well, but its rotation also twists space-time close to its surface, called frame-dragging. Each time Mercury enters this distorted region of space (the gray area), its orbit is slightly altered by a fraction of a degree. This change in Mercury's orbital path is called a perihelion shift.

**Arthur Eddington
(1882-1944)**

In May 1919, a total solar eclipse was scheduled to transit across the southern hemisphere through Brazil. It would cross the equator over the Atlantic, then over the island of Príncipe, in the Gulf of Guinea. To test Einstein's predictions, two teams of astronomers were dispatched to photograph the eclipse. The first team, which would serve as a backup, was led by Frank Dyson and was sent to South America. The primary team, led by Sir Arthur Eddington,[*] traveled to Africa. Eddington took photos of the stars in the region surrounding the eclipse's edge. (Figure 7-10) And when he later measured the degree of deflection of the stars in the photo, he found that their position (i.e., the degree of deviation from their actual positions as caused by the Sun's warping of space-time) matched what general relativity had predicted. We now call the deflection of light around massive objects *gravitational lensing*. (Figure 7-11)

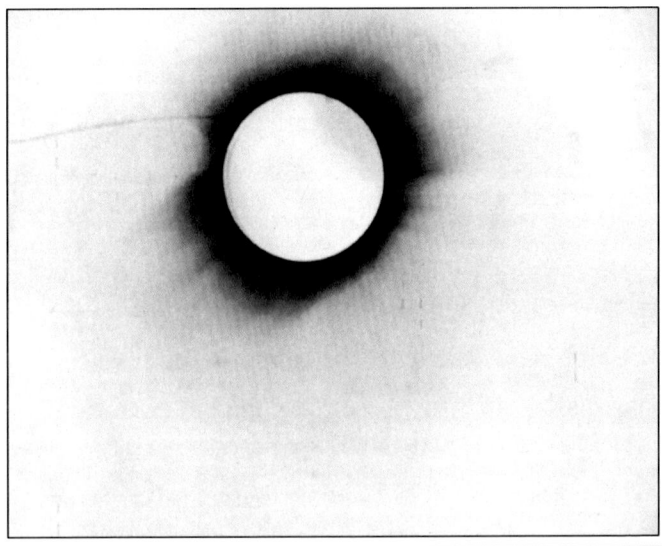

Figure 7-10. A film negative from Eddington's expedition to the island of Príncipe, May 1919.

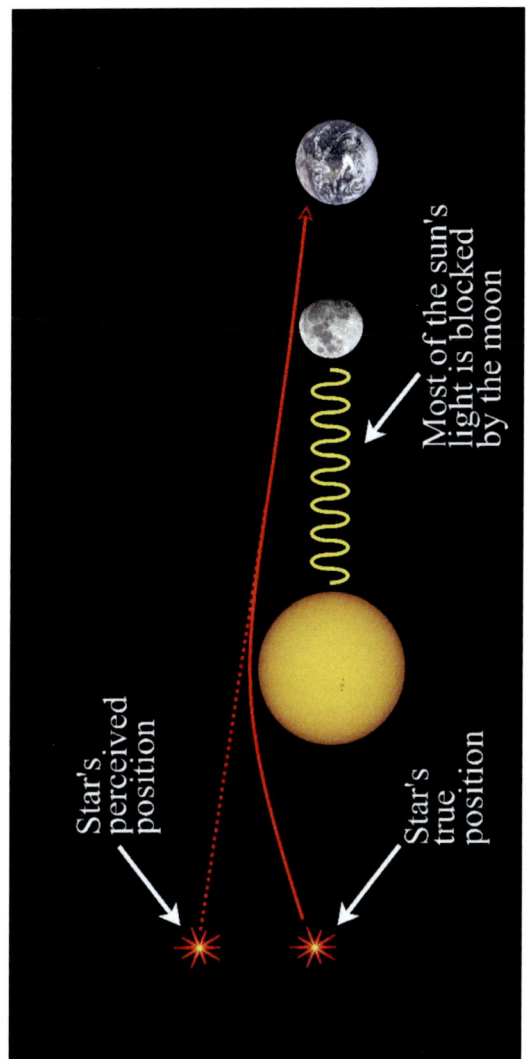

Figure 7-11. Gravitational lensing causes the light from stars located behind the Sun to bend as they pass near the Sun's surface. Normally, this effect can't be observed due to the brightness of the Sun. But when the Moon blocks a large portion of the Sun during an eclipse, stars that should not be visible become observable along the Moon's edge.

Star's perceived position

Most of the sun's light is blocked by the moon

Star's true position

Today, with the aid of more advanced tools, like the Hubble telescope, proof of general relativity via gravitational lensing abounds. (Figure 7-12) It's become a major tool for viewing distant galaxies. Indeed, since the year of Eddington's confirmation, numerous experiments, performed with even greater degrees of precision, have continued to confirm Einstein's theories. And further confirmation came with the first discovery of a black hole.

Karl Schwarzschild (1873-1916)

Within months of Einstein presenting his general theory of relativity, German physicist Karl Schwarzschild made a startling prediction. He stated that the math of general relativity suggested that a star's mass could become so large that distorted space-time could compress it to an infinitely small point. And the result of so much mass concentrated to such a small point was that the space-time around the star would be infinitely distorted. The star would effectively become a black hole against the darkness of space.[*] This mathematical solution became known as the Schwarzschild radius, also called the black hole's *event horizon*. And the first and best-studied black hole was discovered in 1964: Cygnus X-1.[†]

---

[*] Interestingly, the first person to suggest that a star's mass could be so large that not even light could escape was the Rev. John Michell in 1784. Using Newtonian physics, Michell calculated the amount of mass a star would need to possess so that its escape velocity would equal the speed of light. Michell called these stars *dark stars*.

[†] Though they can't be directly observed, there are several ways scientists search for black holes; one way being to look for powerful streams of X-rays. These X-rays do not emerge from the black hole itself since radiation, a form of light, cannot escape a black hole. However, just before matter crosses the black hole's event horizon, it is heated to extraordinary temperatures, which causes it (the matter) to emit radiation. These X-rays flow away from black holes in vast amounts, following the black hole's intense magnetic lines. The second way a black hole is found is by observing an object's motion in orbit of the black hole, typically stars. Therefore, if we see a star orbiting an invisible object at incredible velocities or jets of X-rays, chances are the unseen object is a black hole. Finally, the first photo of a black hole's accretion disk (the material in orbital motion around the black hole) was taken in 2017. (Figure 7-13)

Figure 7-12. As viewed through the Hubble Telescope, at the center of the photo is a bright yellow galaxy. The immense gravity from the yellow galaxy gravitationally lenses the light from a second, more distant, blue galaxy behind it; the more distant galaxy should be obstructed. The yellow galaxy's immense gravity causes the blue galaxy's light to curve around its edges. The result is what is called an Einstein ring. **[Photo courtesy of ESA/Hubble & NASA]**

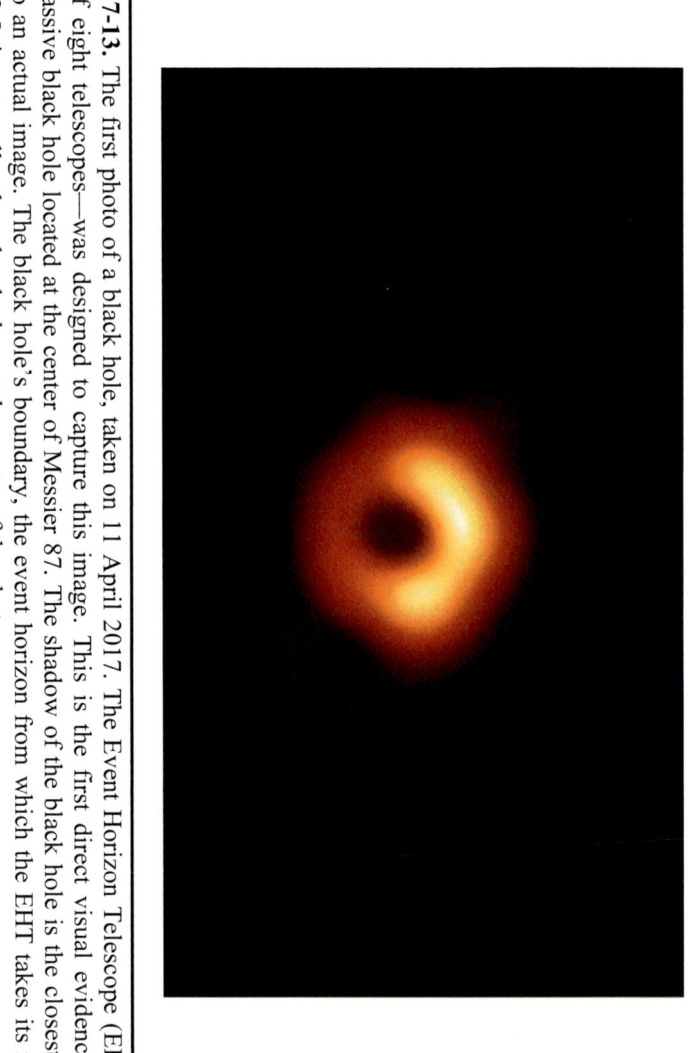

Figure 7-13. The first photo of a black hole, taken on 11 April 2017. The Event Horizon Telescope (EHT)—an array of eight telescopes—was designed to capture this image. This is the first direct visual evidence of the supermassive black hole located at the center of Messier 87. The shadow of the black hole is the closest we can come to an actual image. The black hole's boundary, the event horizon from which the EHT takes its name, is around 2.5 times smaller than the shadow at the center of the photo. Image credit: https://www.eso.org/public/images/eso1907a/

Epilogue

We began with Aristotle's deductive reasoning. Then came Galileo's explanation of relativity and Maxwell's Equations. And finally, we have explored the theories of relativity. If you've followed the logical progression presented, you can confidently say that you understand Einstein's theories of relativity. And you did so without the mathematics usually employed. Yet, with all the effort and experiments of the last 100 years confirming these theories, can it now be said with absolute certainty that they (these theories) are correct?

Another interesting feature of each theory we've considered is that they all fit perfectly into their respective eras–Ptolomy's epicycles or the Copernican revolution, to mention two. For that reason alone, we can confidently say the same about Einstein's theories of relativity; they fit perfectly into our current era. But as society continues to progress, so will our understanding of the universe. Each future discovery will, in turn, allow us to get an even better glimpse at the true nature of the universe. Each bit of knowledge gained will become another piece of an even grander puzzle at which we can only marvel. And every new idea tested will propel us forward toward the next breakthrough. Therefore, as humanity's knowledge improves, so must our theories. It would be arrogant to assume that Einstein's theories are the final answer in describing how the universe operates. In fact, in an ironic twist, Einstein himself sowed the first seed to signal that there was something more to the universe than his theories could explain.

Einstein's theories modeled incredibly well how things behaved on a large scale. But in formulating his theories, he also had to describe the nature of light. Thus, in his paper explaining the Photoelectric Effect, he successfully introduced the idea of wave-particle duality, that light existed as discreet packets, or quanta, of energy. The latter explanation accurately described the behavior of light, which resulted in its immediate acceptance by scientists. However, unanticipated consequences followed.

By accepting wave-particle duality, by the mid-1920s, it had become evident that the microscopic universe followed very different rules. In other words, Einstein's theories are not obeyed on the atomic and subatomic levels – thus, Lorentz was somewhat correct. Indeed, to describe how things operated on this scale, scientists were forced to develop an entirely new branch of physics: *quantum mechanics*. And this new branch of physics was so different from relativity that phenomena that Einstein had resolutely rejected, such as action-at-a-distance, were found to be regular occurrences.[*]

Today, we now have two vastly different theories of how the universe operates. Einstein's theories accurately describe how the universe behaves on a large scale (i.e., planets, stars, and galaxies). While quantum mechanics precisely describes how the universe behaves on a micro-scale (i.e., atomic and subatomic level). And although both have been thoroughly tested and verified to exceptional degrees of precision, no one to date has been able to reconcile these divergent theories. Therefore, just as Aristotle's theories could not account for retrograde motion, the inability of Einstein's theories to account for subatomic behavior indicates that it, too, is incomplete. And while some scientists hold out hope that one day the two theories will be unified, mounting evidence is starting to suggest that this may not be possible.

**Edwin Hubble
(1889-1953)**

Before 1929, the accepted belief among astronomers was that the universe consisted of just the Milky Way galaxy. However, while working at the Mount Wilson Observatory, astronomer Edwin Hubble[†] made an astonishing breakthrough. He was investigating the many fuzzy patches in the night sky that astronomers of his day called nebulae. At the time, it was thought that these were merely luminous clouds of dust and gas located *within* the Milky Way. But to Hubble's surprise, he discovered that these nebulae weren't clouds of dust at all but were other, more distant galaxies. In that instant, the size of the universe expanded

---

[*] It especially disturbed Einstein that subatomic particles engaged in action-at-a-distance, thus violating the speed of light. He dubbed this action "spooky action-at-a-distance," a term that became its unofficial name. Formally, it is called *quantum entanglement*.

[†] Image credit: NASA

exponentially. It went from a single galaxy, which was already immense, to what is estimated today to be more than 2 to 4 *trillion* galaxies.[*] And yet, this was not the most extraordinary aspect of Edwin Hubble's discovery.

As Hubble examined these distant galaxies and measured the amount of Doppler shift their light underwent, he found that their light was being stretched. The light from these distant galaxies was red-shifted, which indicated that the galaxies were moving away from us. (Figure 4-4a) This result was unexpected. Hubble's findings suggested that the universe was not static, as scientists of the era assumed, but was expanding, growing larger. Today, we know that this expansion is even accelerating, a result even more difficult to resolve for two reasons. The first of which was found while studying then-newly discovered clusters of galaxies.

In 1933, a Swiss astronomer named Fritz Zwicky[†] attempted to calculate the mass of the Coma Galaxy Cluster by measuring its rotation rate. Recall that, according to Kepler and Newton, an object's orbital speed depends on the strength of the gravitational force holding it together versus the amount of horizontal velocity the objects possessed. However, as Zwicky measured the rotation rate of the Coma Cluster, he noticed that it was rotating much faster than it should have been. This conclusion meant that the gravitational forces within the cluster were much stronger than first thought.

Fritz Zwicky (1898-1974)

Based on his calculations, Zwicky concluded that there must be more mass than can be seen in the Coma Galaxy Cluster – a lot more! The existence of enough mass to gravitationally influence the rotation of an entire *galaxy cluster* should be visible, and yet, there was nothing to be seen. And that so much matter could not be seen forced Zwicky to come to a strange conclusion. Whatever

[*] https://phys.org/news/2017-01-universe-trillion-galaxies.html
[†] Image credit: Library of Congress

this extra mass was, it didn't appear to interact with light as ordinary matter does; this extra mass only interacted with matter gravitationally. Therefore, Zwicky called this invisible matter "missing mass," which later became known as *dark matter*. Today, as astronomers measure the motion of galaxies, the total amount of dark matter that *must* exist is thought to be at least 5-times greater than all the visible matter in the universe. And the existence of vast amounts of unseen dark matter is a significant problem for general relativity. Why?

The total gravitational force that all the mass in the universe exerts should be drawing all matter together in a process called the "*Big Crunch.*" In fact, with vast amounts of dark matter present, the additional gravity should be hastening the Big Crunch. Yet, instead of galaxies being drawn toward one another, or, at least, the expansion of the universe slowing, the speed at which galaxies are moving away from one another is increasing! Scientists could therefore reach only one conclusion. Along with vast amounts of dark matter, there must also be an even more powerful invisible force of *dark energy* that's pushing the galaxies apart. Think of it: dark energy is causing the universe to expand despite the immense gravity caused by 2 to 4 *trillion* galaxies and 5-times as much dark matter![*]

Yet, neither dark matter nor dark energy fits into any current model of the universe (i.e., the theories of relativity or quantum mechanics). And in recent years, even a key feature of general relativity, the cosmological principle, has been questioned.[†]

---

[*] It's estimated that normal matter such as stars, planets, dust clouds, and galaxies compose only 5% of the universe. Dark matter is estimated to make up 27% of the universe. By comparison, dark energy is estimated to be the remaining 68% of the universe. Ultimately, this means that 95% of the universe, and thus the laws that govern it, are a complete mystery to us.
http://science.nasa.gov/astrophysics/focus-areas/what-is-dark-energy/

[†] To limit the number of solutions to the equations of general relativity, Einstein assumed that matter was evenly distributed throughout the universe. In effect, no structure in the universe should be larger than 1.2 billion light-years across. However, in 2013 astronomers discovered a supercluster of gravitationally bound galaxies, which they named the *Hercules-Corona Borealis Great Wall*. This supercluster measures approximately 10 billion light-years across, 8x larger than general relativity allows.

Consequently, we know for a certainty that Einstein's theories, while wonderfully insightful, are not the final description of the universe. So, what's the next step forward in theoretical physics? Only time will tell.

Still, perhaps the next Aristotle, du Châtelet, or Einstein has already begun to ponder the next great thought experiment. Or maybe the next insight will come from you. Because, as with all the scientists we have discussed, it was not necessarily their mathematical skills that allowed them to discover their small piece of the larger puzzle. Rather, what each had in common was that they all possessed a bit of understanding and a lot of inspiration. More importantly, they were not afraid to challenge what others considered proven facts. It was, therefore, the combination of personality and willingness to invest the time to *ponder* the universe that made their breakthroughs so extraordinary. Remember, for Einstein, it all began at the age of 16 when he asked one simple question: What would it be like to ride a bicycle at the speed of light? Fittingly then, as we all eagerly await the next exciting breakthrough, we can rightly conclude: From the time of Aristotle down to Einstein (and indefinitely into the future), great discoveries are often made by those who are not afraid to imagine.